P9-CRK-679

Polaris student Varvara Andreeva prepares a permafrost core.

Water samples from Siberian streams are colored by the chemical analysis that is in process.

An early autumn frost coats the leaves of a marsh cinquefoil (Potentilla palustris), *a common flower in northern bogs.*

A game trail cuts through the Siberian taiga forest.

The Northeast Science Station's barge is a floating mobile home for scientists working in the Siberian Arctic.

A cat guards the work boots (and teapot) at the warden's house at Zimov's Pleistocene Park.

The Dahurian larch (Larix gmelinii) *is the northernmost-occurring and most cold-tolerant tree in the world. In autumn, its needles turn flame yellow.*

Fallen willow leaves herald the onset of winter in the Siberian Arctic.

THE BIG THAW

ANCIENT CARBON, MODERN SCIENCE, AND A RACE TO SAVE THE WORLD

PHOTOGRAPHY BY **Chris Linder**

Eric Scigliano | With Dr. Robert Max Holmes, Dr. Susan Natali, and Dr. John Schade | EPILOGUE BY **Theodore Roosevelt IV**

BRAIDED RIVER

WITH WOODS HOLE RESEARCH CENTER

CONTENTS

Polaris faculty member John Schade investigates the soil near a Siberian pond.

A fledgling great gray owl explores life outside the nest in the Siberian taiga.

Soil dots the pages of a Polaris scientist's notebook.

Student researcher Erin Seybold takes notes during her stream experiment near the Northeast Science Station in Cherskiy, Siberia.

Trees topple off the edge of an embankment of thawing permafrost along Siberia's Kolyma River.

The roots of ancient plants, possibly dating back to the Pleistocene, are exposed as permafrost thaws along the Kolyma River.

An Evenki reindeer herder gathers his herd in a wooden corral near the Lena River in Siberia.

Autumn brings the first snows to the Siberian Arctic, the world's largest repository of carbon-sequestering permafrost.

The Serge monument overlooking the Panteleikha River near Cherskiy is sacred to the Yakut people. Historically, the posts were used to tether horses and other domesticated animals.

A sandhill crane calls to its mate on the Yukon-Kuskokwim Delta, Alaska.

A mosquito rests on a bloom of hare's-tail cottongrass (Eriophorum vaginatum), *Siberia.*

Snow can come to the taiga forest at any time, even in July.

INTRODUCTION

By Dr. Robert Max Holmes

IT'S NOT HARD TO SEPARATE the optimists from the pessimists when standing on the tarmac in Yakutsk, Siberia. The optimist looks at the aging twin-engine plane our group of twenty-five undergraduate students and scientists is about to board for a four-hour flight to Cherskiy, deep in the Siberian Arctic, and is comforted by its many decades of success at traversing one of Earth's

Polaris students move into "the barge," their floating home for the month of July in Cherskiy, Siberia.

most remote and inhospitable environments. The pessimist looks more closely at the battle-scarred plane and shudders at the countless horrific ways that its good luck streak could end.

I fall somewhere in between and I suspect my friend and colleague Chris Linder does as well. For the past decade we've traveled together over much of the world—from the Amazon and the Congo in the tropics to the Siberian, Canadian, and Alaskan Arctic—all the while doing our best to balance risk and reward as we strive to understand how Earth works and how it is changing. My tools are those of an environmental scientist—sampling equipment, water quality sensors, permafrost drills, and the like—while Chris's are those of a professional photographer. Our partnership has been one of the richest of my career, having allowed me to communicate my science in ways that written research articles alone could never do. Chris's photos also illustrate the human story that is always part of the scientific process, but that is seldom evident in the sterile pages of a journal article or a textbook. And arguably there is no other place on Earth where the scientific story and the human story intersect as powerfully as they do in the Arctic.

My own meandering journey to the Arctic began by chance, soon after I graduated with a BS in zoology. I jumped at the chance to earn a paycheck collecting fisheries data in the Bering Sea for the National Oceanic and Atmospheric Administration, first aboard a 300-foot Korean fishing boat and later aboard a smaller American fishing boat. Interacting with different cultures, weathering frequent storms and the associated monstrous waves, and observing countless fish, whales, seabirds, seals, and other wildlife provided the excitement and adventure I was after. I was hooked on the Arctic. But it wasn't until a decade later, after I earned a MS and a PhD, that I found a way back to my Arctic dream.

OPPOSITE: *The Russian UAZ Bukhanka van is a workhorse on the icy roads near Yakutsk.*

In 1998 I was working as a postdoctoral scientist at the Marine Biological Laboratory in Woods Hole studying nitrogen cycling in estuaries in Massachusetts. As part of that research I developed a method for precisely measuring the concentration of ammonium in waters with very low ammonium concentrations. Some colleagues became interested in implementing that method at the Toolik Field Station on the North Slope of Alaska. They were trying to understand how nitrogen moved through a river system so they could better predict how the ecosystem might change with continued warming, and to do that they needed to measure ammonium very accurately. So I spent two weeks at Toolik, getting the method up and running and assisting on other research projects.

Two years later ammonium again became a central player in my Arctic quest, but it played a very different role this time. I got involved in a project that was compiling discharge and chemistry data from Russian Arctic rivers that had been collected during the Soviet period but were not widely available, even to Russian scientists. Just as a physician analyzes blood chemistry to learn about the health of a patient, we studied river water chemistry to learn about the health of the watersheds. The discharge data provided a wealth of information, but the data on Russian river chemistry—particularly ammonium—was more mysterious. Either something remarkable was going on in these rivers that led to highly unusual results, or the data were bad. After much investigation, which included a trip to Siberia to collect new samples, we determined that the historical chemistry data were deeply flawed. So we decided that the way forward was to begin a project that would collect new samples, essentially starting what would become the long-term record of Arctic river chemistry.

At the start of that project in 2003, which became known as the Arctic Great Rivers Observatory, a colleague and I traveled to the six rivers that were included in the study to meet with collaborators and establish sampling protocols. We went first to the Yukon and Mackenzie Rivers in North America and then to the Ob', Yenisey, Lena, and Kolyma Rivers in the Siberian Arctic.

I'll never forget the feeling of traveling across Russia for the first time, jumping from river to river, in awe of the vast and remote landscape. I knew so little about this region, and few Western scientists even considered working there, yet I was aware that what happened in the Siberian Arctic over the coming years and decades as climate change spurred the thawing of vast areas of permafrost would have global consequences. The sheer adventure of working in such a challenging and awe-inspiring location, combined with the global significance of the carbon-rich permafrost around Cherskiy in the Kolyma region, was the genesis of the idea that grew into the Polaris Project: to lead a group of students and scientists to Cherskiy each year for a research expedition to advance scientific understanding, to train the next generation of Arctic scientists, and to communicate the results to broad public audiences. Cherskiy proved to be an exceptionally interesting yet logistically challenging and expensive location, and after eight remarkable years (2008–2015) we shifted the Polaris expedition to Alaska's Yukon-Kuskokwim Delta, which, like the Kolyma region, was globally significant yet had somehow received relatively little attention from scientists.

The Polaris Project has been successful because of the remarkable people who run it in combination with the extraordinary student participants. John Schade has been absolutely central since 2007 when the Polaris Project was just an idea, and Sue Natali quickly assumed a leadership role after first traveling with the group to Cherskiy in 2012. Together, along with numerous other exceptional colleagues through the

years, we've helped guide the students as they struggle to unlock the secrets of the changing Arctic, and their enthusiasm and idealism has constantly refreshed us. You will meet some of these students, who come from a great diversity of backgrounds, in these pages and hopefully get a sense of the excitement they experience during "moments of discovery"—the ultimate reward of being a scientist.

Our quest to produce a book focusing on the Arctic, told in part through the lens of the Polaris Project, was aided tremendously through collaboration with Eric Scigliano, a journalist with deep interest and a background in the environment and climate change. He was crucial to distilling the story down to its essence and to linking the seemingly diverse threads into a coherent presentation.

As temperatures in the Arctic rapidly rise at twice the global rate, permafrost—carbon-rich frozen soil—is thawing. This is cause for alarm beyond the Arctic, as permafrost contains much more carbon than has ever been released by burning fossil fuels. Thawing releases greenhouse gases into the atmosphere and contributes to rising temperatures worldwide but also causes more localized calamities within the Arctic. As the Arctic ground—frozen for thousands, even tens of thousands of years—thaws and resettles, human infrastructure, including houses and roads, crumbles, along with the ecosystems that are so central to the lives of most Arctic residents. In addition, people living in coastal communities face increased coastal erosion and sea level rise as they watch the ocean encroach on their land, homes, and villages.

Yes, there is cause for alarm. There is more carbon in the permafrost than currently exists in the atmosphere today, and it threatens to be released on an epic scale. However, our hope is that the story from the Arctic,

LEFT, TOP: *Max Holmes, founder and director of the Polaris Project*
LEFT, BOTTOM: *John Schade, Polaris education coordinator*
OPPOSITE: *Sue Natali, Polaris terrestrial research coordinator*

along with the stories of these young scientists who are not afraid to ask questions and who approach research with creativity and resolve, acts as a catalyst for change. We believe the first step is to truly understand the scale and urgency of the problem and then to communicate what we have learned beyond the science community to our neighbors and local communities, to the media and our elected officials. This is core to our work at Woods Hole Research Center and our reason for this book with our publishing partner, Braided River.

As my colleague Sue Natali says, there is a lot of bad news. The good news? Now we know.

OPPOSITE: *The sun rises over a morass of thawed permafrost along the bank of the Kolyma River.*
RIGHT: *In continuous permafrost regions permafrost exists essentially everywhere, whereas in sporadic permafrost regions the majority of the ground is unfrozen.*

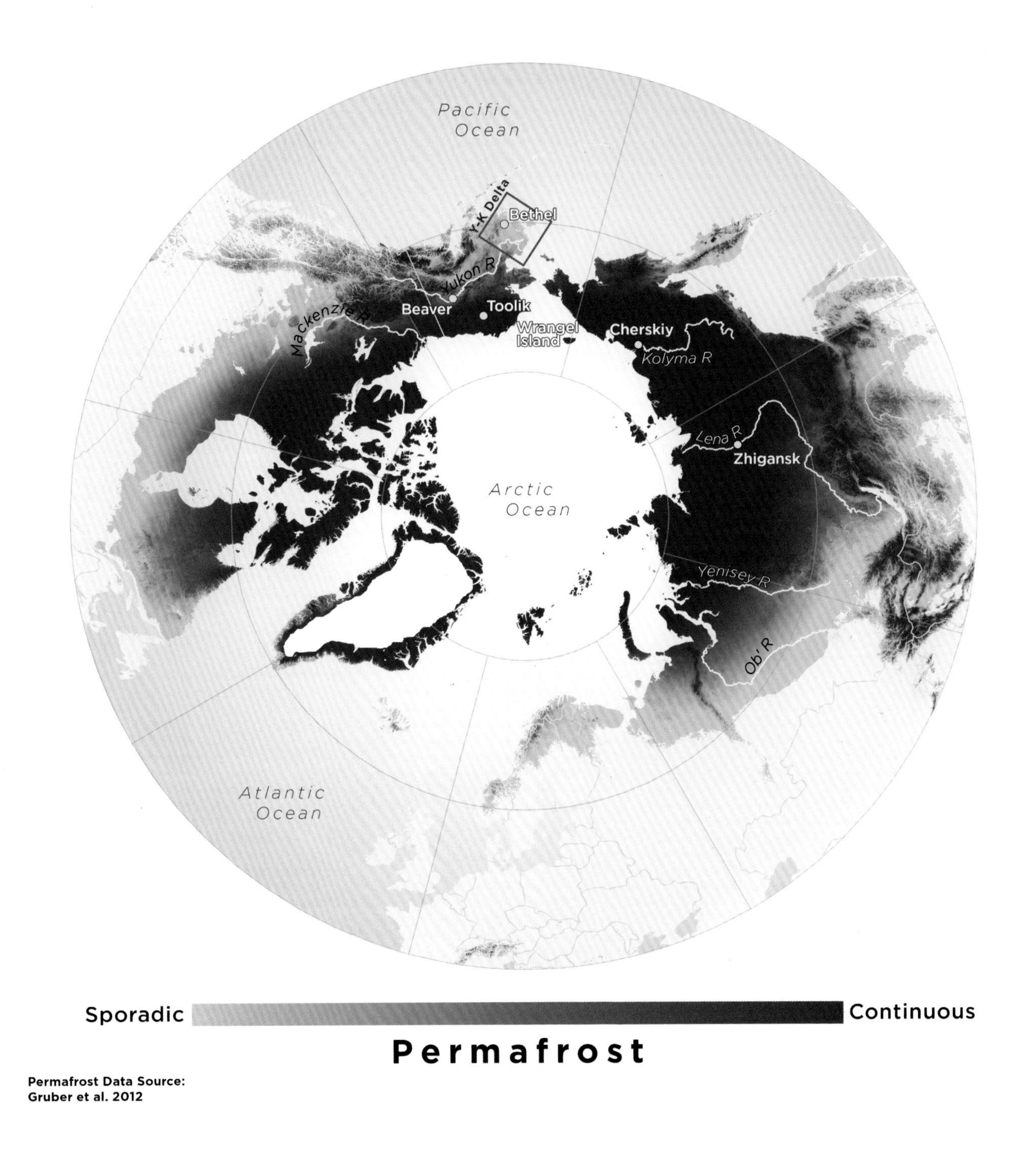

Permafrost Data Source:
Gruber et al. 2012

Natalie Baillargeon measures the depth of the thawed ground.

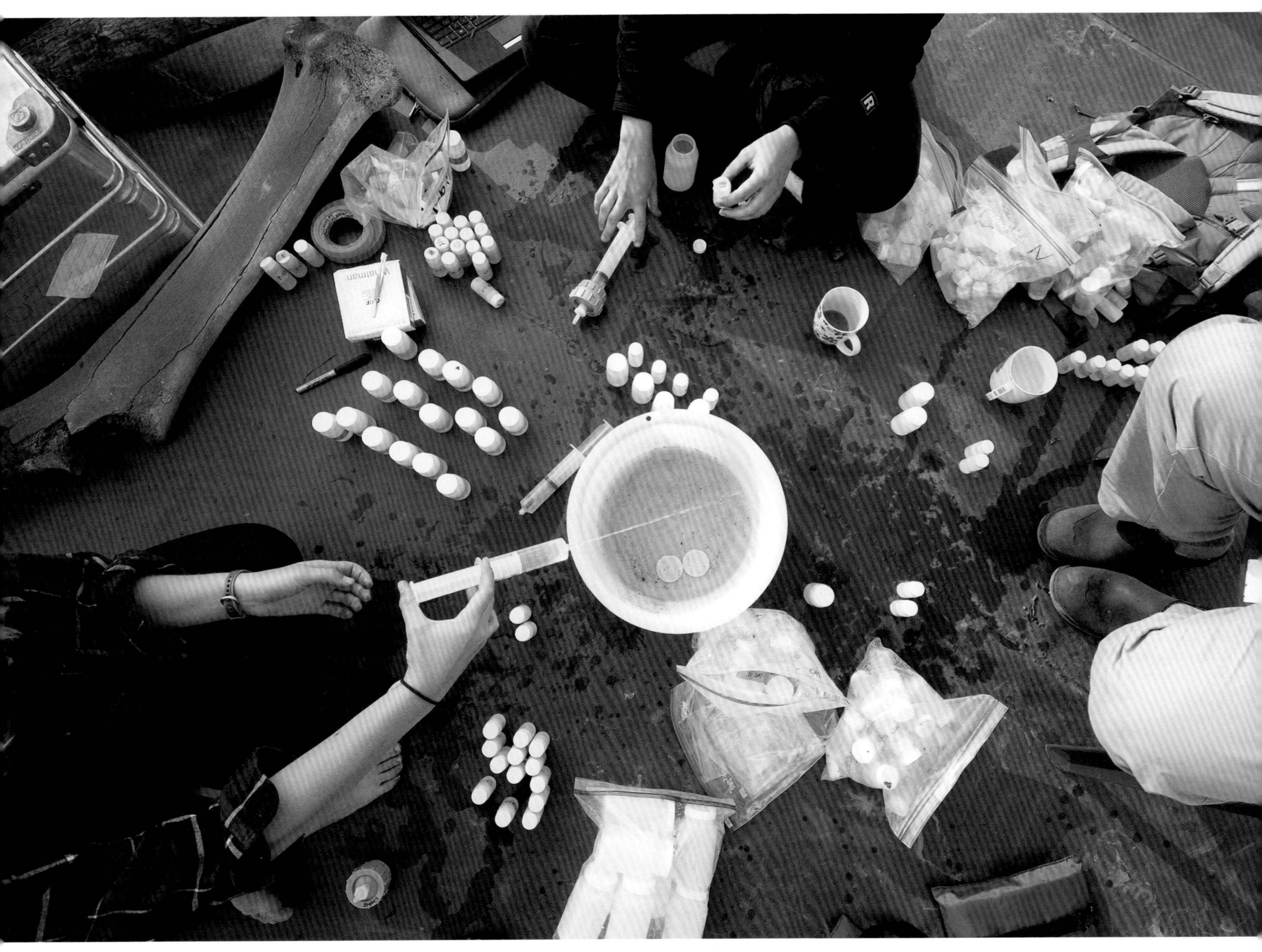

Polaris students and faculty filter water samples on the deck of the Northeast Science Station barge.

Researchers in Siberia contend with endless hordes of biting horseflies and mosquitoes during the summer months.

Scientists take in the view from the wreckage of an old Aeroflot jet in Cherskiy, Siberia.

CHAPTER 1

STRANGERS IN A FROZEN LAND

By Eric Scigliano

ANYA SUSLOVA HAD JUST TURNED 14 when the strangers arrived. They were the first foreigners and the first scientists she had ever seen. Outsiders rarely came to Zhigansk, a northern outpost on Siberia's Lena River that was downgraded from "town" to "rural locality" in 1805 and never regained its earlier status. It couldn't be reached by land in the summer, when the ground turned

Max Holmes returns to camp after a long day in the field, Yukon-Kuskokwim (Y-K) Delta, Alaska.

to marshy, tire-sucking muskeg, or by boat in winter, when temperatures fell as low as -59° Celsius (-75° Fahrenheit) and the Lena froze as hard as the unthawing soil—the permafrost—that lay just a few feet below the stunted taiga forest.

In the warm months, Anya's father skippered a government navigation-aid vessel, keeping the channels clear on the Arctic's second-largest river. Anya rode along with him when she could. That summer, in 2003, a team of Russian and American scientists on a research expedition spanning the Alaskan, Canadian, and Siberian Arctic chartered the boat to take them to various sites on the Lena and its tributaries. There they gathered water samples, chemical evidence of the changes that the vast, little-studied Russian Arctic was undergoing in an age of global climate upheaval. One of the Americans was a tall, rangy fellow named Max Holmes, a specialist in river chemistry from a research laboratory in Woods Hole, Massachusetts.

Max noticed how Anya shadowed him and his colleagues, observing their curious research rituals. He began waving her over to help, and she held the bottles while he filled them. "After a week she'd learned all the procedures," Max recalls with a laugh. "If I skipped a step, she would correct me!"

After two weeks, the researchers had to leave; they would not be able to return to the Lena—a costly, time-consuming trek from Moscow, let alone from Massachusetts—until the next year. But Max had a heap of unused sample bottles and an idea as to how they could still be useful. "Listen," he said to Anya via a translator. "It's really expensive for us to come here, and it would be really helpful if you could continue gathering samples while we're gone."

This was no big deal; Max didn't expect anything to come of it. But when he and his colleagues returned to

LEFT, TOP: *Anya Suslova, Polaris 2008 and 2011 alumnus*

LEFT, BOTTOM: *Erin Seybold collects a water sample from a Siberian stream.*

OPPOSITE: *A network of tributaries feeds the Lena River, one of Siberia's great rivers that empties into the Arctic Ocean.*

Zhigansk the next summer, they were amazed to discover that Anya had continued taking samples along the Lena every two weeks and storing them in a freezer usually dedicated to holding fish and moose meat.

Anya's diligence bore results far beyond the data sets she contributed to. Helping the sojourning scientists set her on the path to her own scientific career. "Max never pushed me to become a scientist," she recounts. "But he gave me two gifts. One was a digital camera. He thought that I might become a great photographer, but I just took selfies and pictures of my girlfriends. The second gift was more important—a subscription to *National Geographic*, in English. It opened my world and got me to learn English. I would learn the vocabulary to translate the articles."

The subscription included a map of the world. "I put it on my wall," says Anya, "and dreamed of *going* places." Sure enough, she did: to university in Yakutsk, Russia, to South Korea as an exchange student, and finally to TERI University in India, where she received her master's degree in climate science and policy. Today she is Max's research assistant at the Woods Hole Research Center (WHRC), one of the world's leading institutions dedicated to investigating climate change. Her father still collects samples along the Lena River.

The ripples from their chance encounter have reached farther yet, across Siberia, Alaska, and the frontiers of earth, aquatic, and climate science and undergraduate science education. "Working with Anya and the other students got me excited about making connections in Russia," Max recalls. Siberia is difficult to get to, let alone work in. "Most Western scientists didn't even try to work in the Russian Arctic, because it seemed like just too daunting a task."

Beyond that, the enthusiasm that young Anya and her friends brought to gathering samples got Max to thinking about the place of young people in science. They, after all, were the ones who would inherit the world that human activity and inaction were now shaping. Might there be a role for them in the quest to understand that world and the consequences of choices being made today? Original research, as opposed to recapitulating solutions to familiar problems, was usually reserved for graduate students and postdoctoral researchers, who were already set on their career paths. Younger students, however, had more freedom to experiment and take chances. Fourteen years old might be a bit young to conduct original research, but why shouldn't talented, motivated undergraduates undertake it?

And thus was planted one of the seeds that would grow into the Polaris Project, an innovative research and educational venture dedicated to twin ambitious goals: to determine what will happen to the vast trove of carbon frozen in Arctic soils as the planet warms and how that will in turn affect the climate, and to recruit, inspire, and train the next generation of Arctic scientists.

Four years later, in 2008, Max Holmes, fellow stream scientist John Schade, and six other colleagues accompanied seven undergraduate students to an even more northerly Siberian research station. There, at the site of an audacious experiment in paleo-ecological restoration called Pleistocene Park, this first Polaris expedition launched a very different experiment that continues to this day.

Since then Holmes and Schade, joined in recent years by ecosystem scientist Sue Natali, British biogeochemist Paul Mann, and other colleagues, have led nearly 100 more rookie researchers to the frontiers of climate change and climate science, first in Siberia and then on an even more remote patch of tundra on Alaska's Yukon-Kuskokwim Delta. With them from the start has been photographer and videographer Chris Linder, who had

OPPOSITE: *Flowering dragonhead* (Dracocephalum palmatum) *blooms on the slopes of Mount Rodinka near Cherskiy.*

already traveled some of the world's wildest rivers with Max, capturing their visual and human sides. The Polaris team understands the importance of communicating its findings beyond the specialists who read scholarly journals. They know that powerful images bring home the realities of vulnerable ecosystems and threatened communities in a way that data alone never can.

A Warming World Crowned by Permafrost

By 2003 the Arctic world that Anya Suslova inhabited and Max Holmes studied had become the focus of urgent scientific interest and environmental concern. What was by then abundantly clear to oceanographers and climate scientists had begun to seep into popular media and public awareness: big changes were underway in the Arctic. Sea ice on the Arctic Ocean was shrinking dramatically, as were the mighty glaciers covering Greenland, and while oceanic oscillations might play some role, only a warming climate could explain the degree of change.

The lands ringing the Arctic Ocean were also changing, but to much less fanfare. Stretching across much of Russia, Canada, and Alaska and smaller areas of several other countries is a vast swath of permafrost, ground that remains perennially frozen beneath a thin surface layer that thaws and refreezes each year. At its northernmost point, on Greenland's north shore, this permafrost belt reaches nearly 84° of latitude, just 420 miles from the North Pole. At its southern margins, the permafrost underlying the Himalaya and Tibetan Plateau lies closer to the equator than to the North Pole, at about the same latitude as Cairo and Houston.

This permafrost belt covers about a quarter of the northern hemisphere's land surface. Some is Arctic desert where little grows, but the share of it that is vegetated—treeless tundra and low taiga

OPPOSITE: *Carbon-rich water fills pockets created by thawing permafrost near Cherskiy.* RIGHT: *A comparison of the amount of carbon in the atmosphere (850 billion tonnes) with the amount in vulnerable carbon pools on land that could end up in the atmosphere*

forest—covers 12.4 percent of Earth's land surface. Tropical rainforests cover only about 7 percent.

Tundra lacks the lush extravagance of Borneo or Amazonia, but it is a land of haunting beauty and dizzying changeability. For two months or so in the Arctic summer, snowy wastes give way to a frenzy of growth, as low-lying plants scramble to catch the round-the-clock sunlight and put out flowers and fruit in order to attract pollen-peddling insects and seed-spreading bears and other foragers. Some 1,700 plant species grow on the tundra (if you count the lichens, which are actually communities of algae and fungi rather than plants). Four hundred are flowering plants that grow so small and low they form a carpet of colors and textures as dense as the finest Persian weaving. In some places their fruits—crowberries, cloudberries, cranberries, and blueberries, each type sweet, tart, and juicy in its own way—grow so thick that you can rake the ground cover as a grizzly does and chomp them by the handful.

From the air, the tundra looks like a uniform plush carpet, but venture out on it and it proves much less regular and more maddening. Sedges form high tussocks that beckon with the promise of solid steps above the doubtful damp tangle below. Step on them, however, and they bend and throw you; if you're lucky your ankle won't twist. Kelly Turner, a 2018 Polaris student researcher, calls it "a vegetative trampoline." Visitors to Alaska's northern rim marvel when Iñupiat hunters say they prefer the cold sunless winters to the warm bright summers. But in summer they find themselves bogged down; in winter they speed over the frozen, snow-covered ground on snowmobiles and ATVs, just as they formerly did on dogsleds.

Viewed from our comfortable warmer niches, this frozen realm may seem as remote and timeless as the surface of the moon. But it *is* changing, with implications that will affect everyone everywhere on Earth.

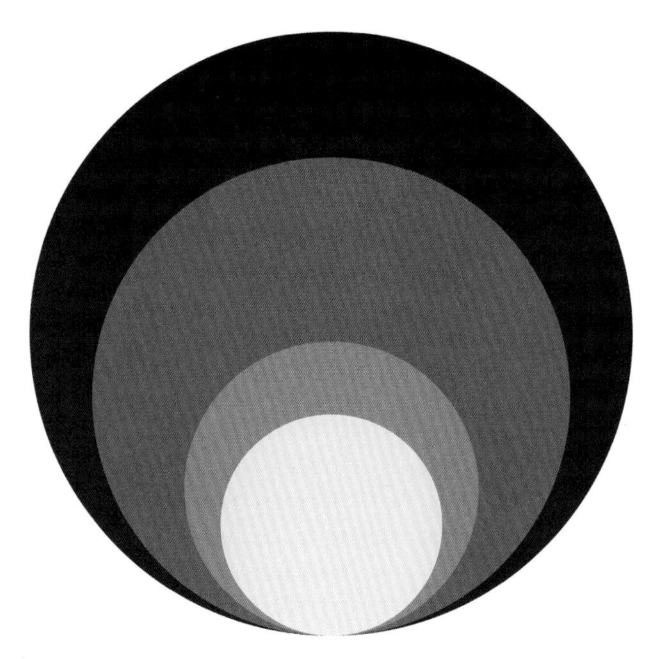

VULNERABLE CARBON POOLS

- *Permafrost, 1,500 billion tonnes*
- *Fossil fuel reserves, 1,200 billion tonnes*
- *Atmosphere, 850 billion tonnes*
- *World's forest biomass, 500 billion tonnes*

Permafrost soils are rich in carbon—the legacy of the grasslands, peatlands, and forests of past epochs, protected by freezing from microbial breakdown. The deep deposits of *yedoma* (the fine-grained soils, also known as *loess*, deposited across Siberia by winds and waters from the south) contain 10 to 30 times as much carbon as ordinary deep mineral soils. In some places, this carbon-rich soil has piled up more than 100 feet deep.

That's an enormous carbon sink, perhaps the biggest on the planet—still intact, though precariously so. The entire permafrost belt holds some 1,500 billion tonnes (metric tons) of carbon—a middling estimate among calculations that range from 1,200 to 1,850 billion tonnes. That's more than the 1,200 billion tonnes of carbon in all the accessible fossil fuels—coal, gas, and oil—that remain (for now) sequestered underground. It's three times as much as all the carbon in all the vegetation on Earth. It's nearly twice as much as the 850 billion tonnes in the atmosphere today.

The general consensus among climate scientists is that another 220 billion tonnes of carbon entering the atmosphere would push average global temperature above preindustrial levels by 2° Celsius (3.6° Fahrenheit), the best-guess threshold before extreme, perhaps irremediable disruptions, from widespread desertification to polar melting and catastrophic sea-level rise, ensue. (At current emission levels, we're on track to far surpass that threshold.) The permafrost ringing the northern hemisphere contains about six times this "emission budget," the amount we can afford to add to the atmosphere if we don't want to upend the planetary systems on which life as we know and love it depends.

Not All Carbon Is Compounded Equally

Permafrost is shot through like fat-marbled beef with polygonal ice wedges, formed when water seeps into cracks in the soil, then freezes and expands. These wedges commonly make up half the volume of Siberian yedoma; they can extend as much as 30 meters (about 100 feet) deep. When the permafrost thaws, this ice melts, forming pools and ponds laden with carbon, nitrogen, and other nutrients, a sort of landscape-scale compost tea. Hungry microbes get to work digesting these nutrients and the rich troves of organic material—peat, plants, whole tree branches, animals large and small—impounded in the soil.

When this digestion (or, as it's more properly called, respiration) occurs in the presence of oxygen, it releases carbon dioxide—CO_2, one carbon atom and two oxygen atoms, the predominant greenhouse gas. When it occurs anaerobically—as in low-oxygen water bodies,

The taiga forest near Cherskiy turns gold as winter approaches.

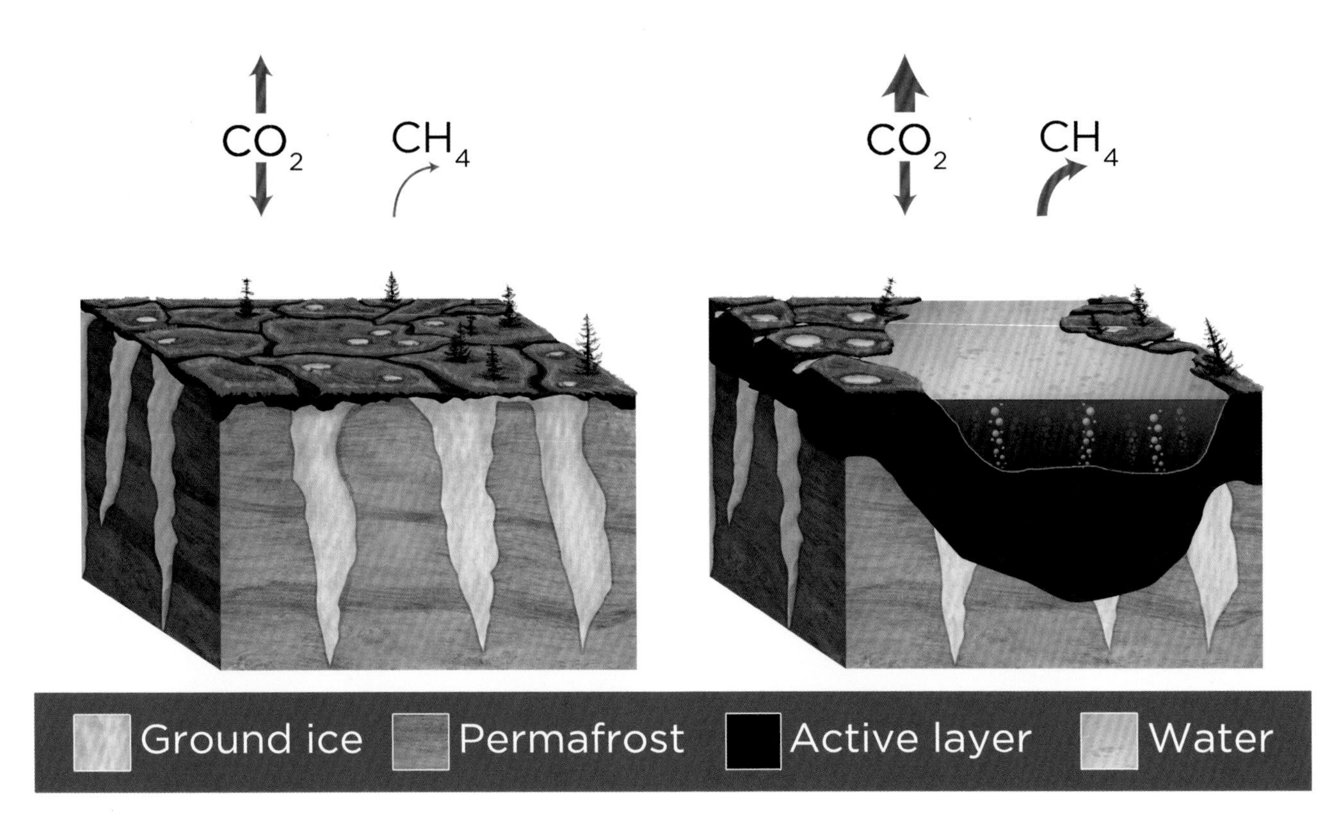

Permafrost thaw can alter the shape of the landscape, in some areas causing abrupt ground collapse, loss of vegetation, and lake formation. These changes in turn have important implications for greenhouse gas emissions, shown here as arrows that widen as emissions increase. More methane (CH_4) is produced and released in saturated areas, such as lakes, and vegetation losses will also reduce carbon removal from the atmosphere.

sodden soil, and the guts of cattle and other ruminant animals—it tends to release both carbon dioxide and methane, CH_4, one carbon and four hydrogen atoms, a scarcer but much more potent greenhouse gas. Molecule for molecule over a 100-year period, methane traps about 30 times as much heat from the sun as carbon dioxide. Methane's relative impact is much greater in the short run, because it lasts less than a tenth as long as carbon dioxide does in the atmosphere. Within about a decade of being released, CH_4 combines with atmospheric ozone, O_3, to form CO_2 and water vapor, both greenhouse gases. The water vapor is soon funneled into clouds, rain, and snow, but the carbon dioxide lingers much longer—100 to 200 years for the most part, though a small share can last hundreds of thousands of years, trapping heat all the while. And thus the mischievous methane molecule gains a second climate-altering lease on life.

These chemical dynamics are just a few in a whole suite of processes that will determine how much of its carbon the permafrost repository releases, how fast, and in what chemical form. Other processes can moderate and slow or magnify and speed the warming effect. "Positive feedback loop" is a once-esoteric term that's gaining wider currency: A makes B happen, B makes more A happen, and more A in turn makes more B happen—a repeating loop. When more heat reaches the earth's surface, it thaws more permafrost, which

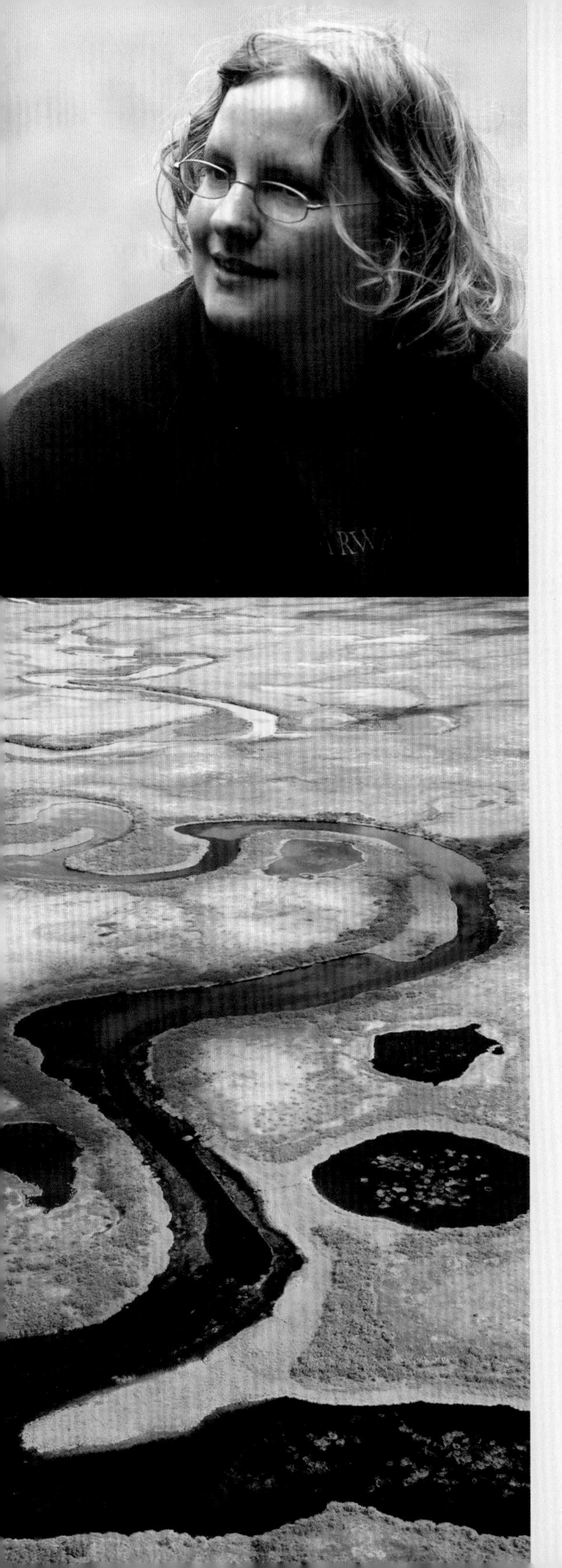

CLAIRE GRIFFIN

When Claire Griffin arrived in Cherskiy as a Polaris student in 2009, she had never been to the Arctic and had never done field- or even lab work, beyond counting tree rings in a forest ecology lab at Clark University in Massachusetts. But the research she did in Cherskiy produced a paper in the *Journal of Geophysical Research* and three conference presentations.

Griffin set out to determine whether the colors—or, rather, shades of the ubiquitous tea color—of Siberian lakes and rivers captured in satellite imagery could be used to gauge how much dissolved organic carbon they contained. Such a tool would be a force multiplier for environmental monitoring, she explained at the time. "If we can do that, we can look at more lakes than we can reach directly," she said. "I hope to look at many, many more lakes."

She shifted her focus to rivers across the Arctic in her graduate studies and joined the Polaris Project again in 2013. Mapping organic matter "involved a lot of processing and boring atmospheric correction," she explains. But she recalled John Schade's advice—"Every type of science involves some type of tedium. Find the tedium you can tolerate or even enjoy and do that"—and persisted, incorporating reams of data collected by Polaris and the Woods Hole Research Center's Arctic Great Rivers Observatory.

Today she is using the techniques she refined in Siberia to investigate water quality and algal blooms in lakes across Michigan, Wisconsin, and Minnesota—which, like Siberia, have more lakes than anyone can get to directly. She wants to return to work in the Arctic and pass along what she gained from Polaris to other undergraduates. A commitment to science means not just research but outreach, she explains. "If we want people to be excited about science, then giving them access to that science will spread the knowledge and enthusiasm. Being able to *see* [what's happening in places like Pleistocene Park and the Yukon-Kuskokwim Delta] is what will get people invested in climate change.

"If we want people to be excited about science, then giving them access to that science will spread the knowledge and enthusiasm."

You need to not just say 'This much carbon is being put into the atmosphere' but 'I went to these places and I saw these changes, and here's what's going on.' By helping students experience that change, I believe we can spread the message even more widely.

"I don't want to be a pessimist, but it's a really challenging situation. Arctic permafrost is a prime example of a positive feedback loop, of how we're changing the environment, making the situation worse. It's something we have to be able to account for in our climate models.

By helping students experience that change, I believe we can spread the message even more widely.

"There are a lot of ways that change in the Arctic is going to affect things on a global scale. And the Arctic is the canary in the coal mine—it's changing faster than any place in the world. Permafrost thaw is not the single thing that's going to push us into a new climate regime. But it's a big piece of the puzzle."

releases more carbon, which causes more heat to reach the surface, which . . .

Another factor that can strongly influence this greenhouse loop is *albedo*, the share of radiation—in this case, solar radiation—that a surface reflects. You can experience albedo by standing in the hot sun and changing from a white shirt to a black one; the black shirt, with its low albedo, gets much warmer than the high-albedo white shirt. Snow and ice, which are highly reflective, have high albedos. Open water has a much lower albedo, and so absorbs more heat.

The process plays out on land as well. Trees and shrubs tend to be darker—with lower albedos—than the sedges and lichen of the tundra. The black char left by fires has a lower albedo yet. The dark twigs and trunks of tall shrubs and trees stick out from snow that would cover shorter vegetation, absorbing more heat and in turn causing the snow to melt faster.

These processes, like most processes in nature, are not one-way streets. If shrubs enable snowdrifts to pile up higher, the snow cover may last longer into spring, extending its albedo effect. But these higher drifts compound the insulating effect of the snow (which, because it contains so much air, is a highly efficient insulator), conserving surface warmth and delaying the winter freeze.

As with heat and light, so with moisture. Will the Arctic get drier or wetter as it warms? "My answer to that question is yes," says Holmes—both, in different ways, to different degrees, in different places at different times of year. Calculating the net impact means balancing a host of contrary trends, and, he cautions, "the answer never will be definite—it depends on a lot of variables."

Consider some of these variables. Freezing seals off soil, preventing water from draining into it or out of it. Drier soil stays dry, saturated soil stays wet, and water

OPPOSITE: *Ice wedges and mounds of thawing Pleistocene-era permafrost soil form an otherworldly landscape at the exposed riverbank of Duvannyi Yar on the Kolyma River, Siberia.*

OPPOSITE: *Exposed permafrost along the banks of the Kolyma River*
RIGHT: *Stringlike roots extrude from the thawing permafrost at Duvannyi Yar.*

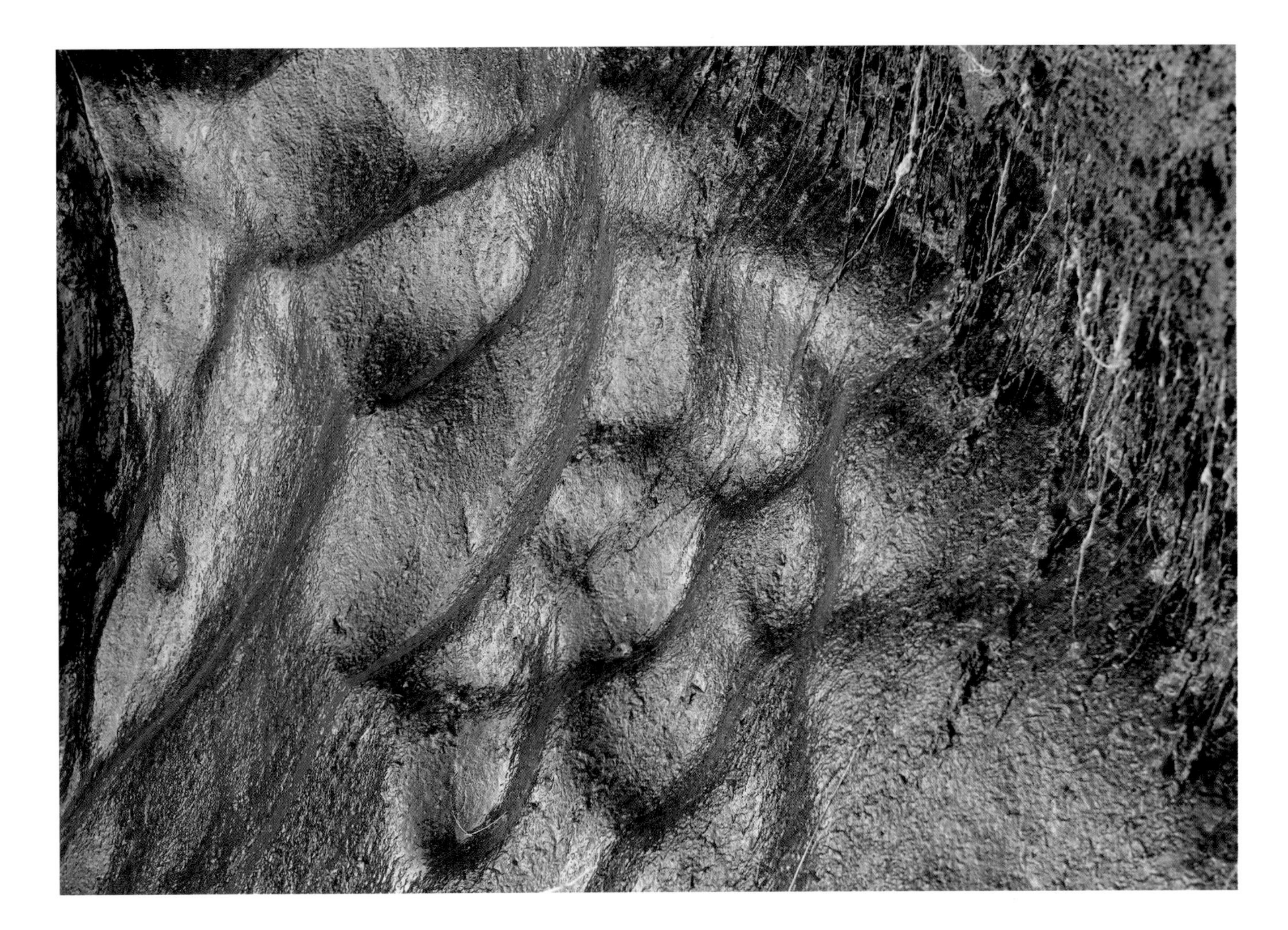

that penetrates the unfrozen surface runs off the frozen layer below. As permafrost thaws, it may release or absorb runoff, depending on terrain—a process that may itself be altered by thawing, as ice melts and the ground subsides. Upland permafrost tends to drain and dry out as it thaws, releasing large quantities of carbon dioxide. Lowland permafrost and upland depressions tend to soak up and hold runoff as they thaw, becoming wetter. Microbial respiration proceeds more slowly in these saturated soils, so they release less carbon. But they release it in the much more greenhouse-effective form of methane.

How much of the permafrost's carbon lode will escape as carbon dioxide and how much as methane? Answering that deceptively simple-sounding question will entail unraveling a complex web of chemical, thermal, atmospheric, biological, and hydrological factors.

Blueberries and irises grow in profusion in the short, intense Siberian summer.

The hills along the Kolyma River, at the eastern edge of Siberia, glow purple under the midnight sun.

As the formerly firm permafrost soil beneath them thaws, larch trees fall into a Siberian lake while methane bubbles up from the lake's bottom.

Wildflowers bloom around a baydzerakh, or mound of thawing permafrost soil, on the Kolyma River.

CHAPTER 2

FIRE ON THE TUNDRA, REVELATIONS IN THE RIVERS

WARMING INJECTS ANOTHER WILD CARD into the already complicated dry/wet-carbon dioxide/methane-albedo/insulation equation. Hotter summers produce more thunderstorms—and more lightning. This, together with drier vegetation, has brought something that sounds improbable in a frozen land: an epidemic of wildfire, not just in boreal forests but on the tundra itself. Dry shrubs and

Fire transforms the tundra landscape of the Y-K Delta, exposing the charred ground to faster warming.

OPPOSITE: *Near Cherskiy, the taiga forest is composed mainly of Dahurian larch* (Larix gmelinii).
RIGHT: *A fogbow, or colorless rainbow, forms over a lake in the Y-K Delta, Alaska.*

grasses are the tinder, lightning the match. In 2015, 726 square kilometers (280 square miles) burned on the Yukon-Kuskokwim (Y-K) Delta in western Alaska, where the Polaris Project now works. Just 477 kilometers (184 square miles) burned in all the preceding 74 years combined.

Fire is already transforming the tundra landscape in frontline regions like the Y-K Delta. Shrubs thrive in soils disturbed by fire and other upheavals that remove competitors like deep-rooted sedges and grass. They have rapidly occupied burned-out patches in western Alaska and on the North Slope and are encroaching on vast swaths of warming tundra.

Unfortunately, it can be devilishly difficult to observe and document this encroachment in real time. Like a child's growth, it proceeds continuously and imperceptibly, unless you have a benchmark for comparison.

Such a benchmark appeared unexpectedly about 20 years ago: the discovery of thousands of aerial photographs taken across the North Slope in 1944 by the US Navy, the first step toward finding the new sources of petroleum that the Allies thought they'd need to sustain the global war effort.

Starting in 2000, researchers retraced some of these transits by helicopter, taking matching photos at 200 sites covering about 194,000 square kilometers (75,000 square miles). These before-and-after images, coupled with ground observations, told a striking tale: the warming climate was promoting a dramatic shift in vegetation. What had been small shrubs were now tall and bushy, and new patches of willow, birch, and alder, the "shock troops" of landscape change, had popped up in areas where no shrubs tall enough to be detected (about 50 centimeters, or 20 inches) had appeared before. As shrubs grew tall they encouraged the growth of outward-spreading concentric rings of smaller shrubs, each shielded from the elements by its predecessors.

Over the same 56-year period, taiga forest—which requires more water and warmer conditions than tundra does—had encroached on an estimated 11,700 square kilometers (about 4,500 square miles) along the southern margin of the Alaskan tundra. The same processes have been observed, on an even larger scale, in Siberia.

A Planetary Blood Workup

It was rivers, together with a craving for adventure in distant places, that drew Max Holmes to the Arctic. And it is rivers that now provide an essential tool for tracing the changes transpiring in the lands they drain, the seas they feed, and the climate that has succored and now threatens life on Earth.

Holmes grew up in Traverse City and Ann Arbor, amid Michigan's lacework of streams, lakes, ponds, and wetlands. As soon as he could get his feet wet he conceived an enduring passion for fishing and whatever else would let him poke about in streams. "I loved to look under rocks and check out bugs," he recalls. "That's what drew me to science."

Perhaps he intuited then what he would come to realize as he pursued his scientific education and career: the vital role that streams and rivers play as collectors and transmitters of chemicals, organisms, and information. "I'm interested in river chemistry for the same reason a physician is interested in blood chemistry," Holmes explains. Physicians examine blood samples to discern the health of the entire organism. A single sample can reveal something acutely wrong; a high blood glucose level suggests diabetes, and soaring creatinine points to kidney failure. But even if all measurements are within normal ranges, multiple samples over time can reveal significant changes—diseases in the making.

Likewise for the rivers, which was why young Anya Suslova's steadfast sampling on the Lena was so welcome. It was also why Holmes and his colleagues were so excited to discover a long-forbidden trove of discharge volume and chemistry data for Siberia's rivers, which came to light following the collapse of the Soviet Union. (It wasn't just official paranoia that kept these data locked away, even from Russian scientists; river flows had strategic significance. Freshwater infusions affect seawater's density and acoustics, hence its ability to conceal submarines. In a 1968 Russian movie called *The Secret Agent's Blunder*, a Western spy covertly collects water samples downstream from a Soviet military base.)

"The beauty of rivers is they *integrate*," says Holmes. Measuring soil chemistry—from carbon content to trace minerals like mercury—is difficult and time-consuming, especially across the roadless expanses of a continent-sized region like Siberia. Readings can vary greatly even between sites a few feet apart. Sample at a river's mouth, however, and you get a distillation of all the lands it drains—a broad-brush view, to be sure, but useful for tracking changes, and you can then travel up the river's course, sampling its tributaries, for an increasingly precise, fine-grained measure.

River flow is especially telling in the Arctic, not just for what it says about the lands upstream but for how it affects the ocean waters downstream. The Arctic Ocean, smallest and shallowest of the world's oceans, contains only about 1 percent of the seas' total volume. But it receives about 10 percent of all the freshwater flowing

Travis Drake (left) and Blaize Denfeld collect water samples from a Siberian lake.

into them. One researcher likes to call it "the Arctic Estuary."

Six great rivers deliver most of that freshwater: the Lena, Yenisey, Ob', and Kolyma in Siberia, the Mackenzie in Canada, and, running from Canada's Yukon Territory across Alaska to the Bering Sea, the Yukon, which shares its wide delta with the Kuskokwim River to the south. Each of the first three is comparable in annual discharge to North America's largest river, the Mississippi. All function as biochemical gauges, revealing the changes unfolding in their watersheds.

Soviet scientists began reading these gauges long before their Western counterparts caught on. They started measuring the discharges of their Arctic rivers in the mid-1930s and began measuring their chemistry in the 1960s and 1970s. By contrast, Yukon and Mackenzie discharges weren't monitored until the 1970s. In the 1990s, Holmes worked with Russian and American

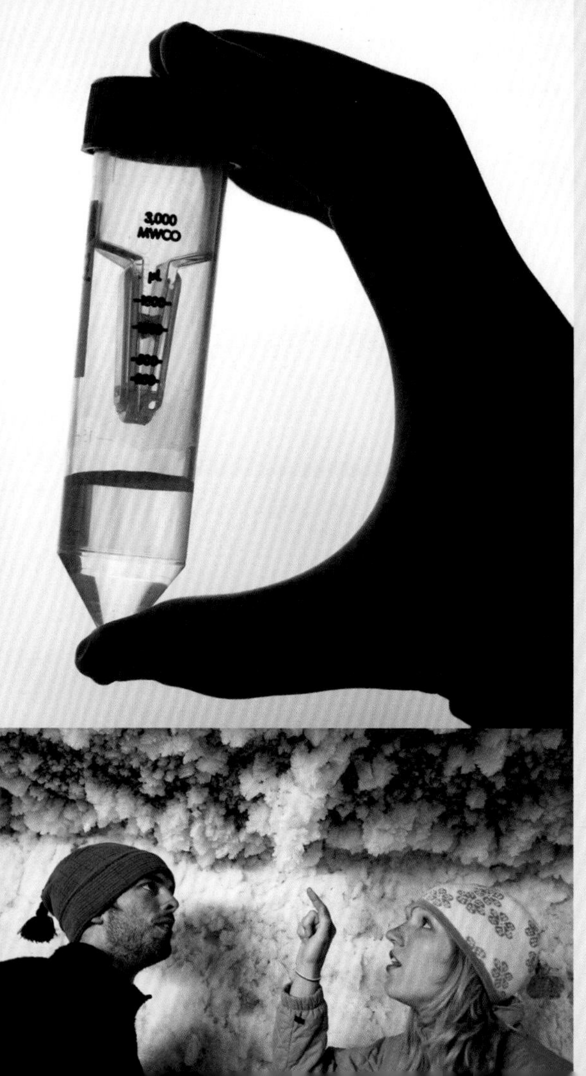

TRAVIS DRAKE

Serendipity played a part in Travis Drake's joining the second Polaris expedition, in 2009, and discovering his calling as a biogeochemist. He was majoring in geology at Carleton College in Minnesota "because it gave me the opportunity to be outdoors." Biogeochemistry—the interactions between earth systems and chemical and biological processes—was not much on offer there. But he signed up for a course on Arctic systems taught by a visiting professor named Max Holmes, who announced that slots were available on an expedition to Siberia—and never looked back.

"My focus on carbon biogeochemistry was codified with the Polaris Project," Drake says. "It was a pretty easy choice—it seemed like the most important elemental cycle in the Arctic and even globally. The potential that the Arctic has to contribute carbon emissions, the fact that it was the sleeping giant up north—I knew I wanted to be involved."

He still recalls "how hugely informative it was for me to actually sit with a real ecologist"—John Schade—"who would go out in the field, bring samples back, and analyze them. That was the first time I saw how the sausage was made. That hooked me."

He returned with the 2010 Polaris team to continue his research into the ways Siberian streams process the key nutrients nitrogen and phosphorus. (His findings suggested that small Arctic streams could become important sources of carbon dioxide emissions as permafrost thaws and releases more nutrients.) He returned again as a graduate student to research further and help the new undergraduates with their projects. Now, as a doctoral candidate at Florida State University, he's using the measurement methods he honed in his Polaris summers to investigate how deforestation and land conversion to agriculture affect the release

"The potential that the Arctic has to contribute carbon emissions, the fact that it was the sleeping giant up north—I knew I wanted to be involved."

of carbon into rivers in the Congo basin.

To focus on your research, you can't obsess over the dire global context.

After that? Drake expects to continue his Congo research as a postdoc at the Swiss Federal Institute of Technology in Zurich. "Eventually I would love to come home" to the States, he says, "but I don't know what the prospects will be. They seem to be hiring more climate researchers and professors in Europe." He got a glimpse into how political circumstances can influence science funding during a stint researching outgassing from rivers with the US Geological Survey in Colorado. "The whole duration I was there they had a hiring freeze. It was really difficult to bring on young researchers."

Drake notes a couple of ironies to pursuing climate research "as a human who cares about humanity all over the world." To focus on your research, you can't obsess over the dire global context. "I don't know if I'm just good at compartmentalizing, but I've found it kind of easy to take a more detached position as a researcher. Advocacy and policy seem way more frustrating and a lot less rewarding—the direction we should be moving seems really obvious."

And from a climate view, "science is kind of a double-edged sword. A lot of scientific work is resource- and energy-intensive. I have to fly around the world to gather my samples. The technology I use to store and process them has a pretty big carbon footprint. You hope that the work you're doing is somehow worth all the emissions."

Further complicating the picture, "the insights I got working in Congo are quite different from those in Siberia. Millions of people in Africa desperately need access to energy, clean water, medicine, and reliable food supplies. That's how we get people to the transition where they stop having so many children. But how do we do that without losing the remaining rainforest, destabilizing all the carbon that's buried there, and burning insane amounts of coal and gas?"

colleagues to gather, digitize, and analyze this scientific treasure trove and build a baseline for tracking future change. They found a trend that evinced significant climate impacts already and potential further impacts on both climatic and oceanic systems. Since 1936 total discharges from the six largest Eurasian rivers had grown by 7 percent, indicating both precipitation and temperatures had been rising. If it continues growing, this freshwater boost to the Arctic Ocean could destabilize the North Atlantic circulation systems that draw warm water from the south—the famous Gulf Stream—and that make Europe's climate so different from Siberia's and Labrador's.

Holmes and his colleagues found wide discrepancies in the chemical data recorded by the Soviets, however. In 2000, under the auspices of the Marine Biological Laboratory (also in Woods Hole), where he then worked, they came to Russia to take new samples for comparison. Those data confirmed that the old chemical data were deeply flawed, so the laboratory launched an ambitious new project sampling the waters of all the great Arctic rivers at multiple times of year—a project later taken over by the Woods Hole Research Center, which Holmes moved to, and dubbed the Arctic Great Rivers Observatory. It was that project that brought him to Zhigansk in 2003.

A Cascade of Effects

Like Max Holmes, British biogeochemist Paul Mann was fascinated by the tale the rivers tell and drawn to the Arctic by the allure of the unknown. "I signed up for the adventure," he says—and got more than he expected. In 2010, fresh out of graduate school in England, Mann applied online for a job as a postdoctoral researcher in the Global Rivers Observatory, another project codirected by Holmes at the Woods Hole Research Center that was monitoring biochemical changes in several of the world's most important rivers. "We've got all these projects going on," Holmes told him. "Want to go to the Congo?"

Two weeks later Mann was there with Holmes, surveying the chemistry of the small streams feeding Africa's mightiest river. A few weeks after that he was off to Siberia's Kolyma River to capture the carbon-rich water surging from the springtime ice breakup and help set up a lab for that summer's Polaris contingent: "I remember a big old Russian bus coming down the dirt road, full of eager, fresh-eyed students. . . . The enthusiasm and excitement just take you in!"

Since then he's undertaken a flurry of research projects in Siberia, Alaska, and the Canadian Arctic, publishing widely all the while. But he and other busy scientists still make time each summer to join the Polaris Project, even though mentoring undergraduates is hardly the usual fast track to professional success.

Still boyish in aspect and bubbling with excitement, Mann finds Polaris gratifying not only as a teacher but also as an alumnus and perpetual student. There's the challenge and the satisfaction of, as he puts it, "teaching undergrads to be fantastic Arctic scientists—I'm convinced some of them will make tomorrow's breakthroughs." At the same time, he adds, "I still look at what Sue, John, and Max are doing and learn as I go along."

And there's the great Arctic puzzle to be solved: what impacts will cascade through the terrestrial, alluvial, oceanic, and atmospheric systems as the frozen ground warms? The carbon-saturated water released doesn't necessarily flow straight into streams and down to the sea. Much of it collects in ponds and lakes called "hot spots"; thawing ice wedges leave depressions that fill to form more hot spots. The term is particularly apt; these nutrient-rich ponds are hotbeds of microbial activity

OPPOSITE: *Crepuscular rays filter through the fog-enshrouded taiga at 4 a.m., Schuch'ye Lake in the Siberian Arctic.*
RIGHT, TOP: *Paul Mann and Max Holmes steer their boat into a small Siberian stream. The river is cloudy with dissolved carbon leached from thawing permafrost.*
RIGHT, BOTTOM: *Mann watches a floatplane take off in the Y-K Delta.*

and prolific generators of greenhouse gases—in particular of potent methane.

Past modeling of these methane emissions has tended to focus on larger ponds and lakes, in part because they're easier to identify in satellite images. But Margaret Powell, a student at Harvard University who joined the 2018 Polaris expedition, found that this approach may miss a significant share of these emissions. She measured dissolved methane at the surface and found that it was "one or two orders of magnitude greater" in small ponds than in lakes. She suspects that's because the organic material that microbes turn into methane enters water bodies from the sides, and small ponds have longer perimeters—more side—relative to surface area. This would mean that total Arctic methane releases may be even greater than standard models have suggested.

Fire raises the stakes yet further. Tundra fire doesn't merely lower the ground's albedo; it transforms the terrain in ways that promote further warming and thawing.

For starters, tundra fires burn not just the vegetation above the ground but the fluffy, insulating upper layer of organic soil, as much as a foot deep. With this cover removed, summer heat penetrates deeper, thawing the permafrost below. In 2012 Sue Natali, Max Holmes's successor as Polaris expedition leader, and collaborator Heather Alexander conducted experimental burns in Siberian larch forests. Two years later, thawing still extended significantly deeper in a plot where 75-plus percent of the organic soil layer had been burned away than in an unburned control plot and in plots where much smaller shares of that insulating layer had burned.

"We think of fire as 'trees burn, soils burn,'" says Natali. "But there's a whole cascade of effects." In addition to removing insulating ground cover and decreasing albedo, the burning changes the soils' hydrology. Natali had hoped to quantify these changes during the

MIA ARVIZU

"I don't know how I came from there!" Mia Arvizu says with a laugh, recalling how she grew up in Bakersfield, a mecca for oil drilling, almond growing, country music, and conservative politics in California's San Joaquin Valley. "At OSU [Oregon State University, where she now studies] we're the Beavers. This guy comes out at games in a beaver costume. But in my high school it was the Drillers, and the mascot was this big guy with muscles and overalls and a helmet."

Neither love of nature nor a passion for science ran in Arvizu's family. Somehow she found both, and they led her to the Polaris Project and the Arctic. "I don't know why, but I just knew that I wanted to do environmental science. Maybe it was some connection from hiking and going to national parks. I actually led this ecology club in high school, sending people out to Yosemite and Big Sur and learning about the environment and how we're degrading it so much. I wanted to do something to help, or at least understand the processes that are affecting it so we can fix it."

Still, until she stumbled on the Polaris Project, she never expected to go to, much less do research in, the Arctic. "It was just a trip to Alaska, doing science. It sounded great." Great it was, but also daunting. "There was one night before an online meeting where we were supposed to present our research questions. I was freaking out, feeling like, I don't know, my question's not good enough. I just started reading papers. I got so stressed that I had to stop thinking about it. So I took a shower, and it came to me like that."

The project that came to her was ambitious: to investigate the levels of nitrogen—in particular the nitrogen compound ammonia, a potent fertilizer—in burned and unburned upland and lowland soils, and their relationship to methane and carbon dioxide emissions. She found much more total nitrogen and ammonium in the low, wetland soils, together with higher carbon dioxide emissions and much higher methane emissions. This suggests that, although nitrogen can be toxic

> *"I wanted to do something to help, or at least understand the processes that are affecting it so we can fix it."*

to microbes in some circumstances, it's supercharging the microbes that produce methane in thawing permafrost.

These initial findings suggest more questions, and Arvizu feels the tug to pursue them. "I would definitely enjoy doing this again. The tundra is such an interesting place to explore because it's so dynamic. It's always changing."

"We have all this data, and if we only used it, the world would be a better place. But it takes a long time before anybody does anything with it."

But much as she enjoys it, she has reservations about continuing in research. "Before I came here I was thinking about switching to ecological engineering. I like science, but I was having a hard time finding value in all the data we're collecting. We have all this data, and if we only used it, the world would be a better place. But it takes a long time before anybody does anything with it. You have to hope politicians or engineers find it and implement some policy or create a new design. It seems frustrating to put in all this work and only twelve people read your paper."

Given these reservations, Arvizu has also considered working on policy or starting a nonprofit. "I definitely would not be a straight engineer. The ecological engineering major at my school has a really big emphasis on science. It's all about taking natural systems and implementing them in your design. If more engineers included the environment in their designs, the world would be a better place."

2018 Polaris expedition: "Unfortunately, foxes repeatedly destroyed my moisture probes, so I don't have much moisture data from the Y-K Delta." But it appears that with the fluffy, absorbent upper soil gone and few plants left to take up water, rainwater quickly saturates what porous soil is left above the permafrost and the dense, compacted soil below. Excess water runs off, laden with carbon, to pool in those methanogenic hot spots.

"These hot spots become saturated with gases and nutrients and spew vast amounts of greenhouse gases," says Paul Mann. "The question I'm interested in is, How will this affect the atmosphere?" The answer depends in large part on how much of the carbon escapes as carbon dioxide and how much as methane. "The early indication is that there's much more methane because of the 2015 fire" in the Yukon-Kuskokwim Delta.

Pursuing their individual research projects, Polaris's undergraduate participants are gathering a range of data that affirm, amplify, and help to explain that indication. On the 2017 expedition, Jordan Jimmie, now a graduate student at the University of Montana, sampled the waters of 16 streams in the Y-K Delta as they flowed through tundra that burned in 2015 and through unburned areas. Samples from the burned reaches showed dramatically higher levels of three key nutrients: nitrate, ammonium, and phosphate. They had twice as much phosphate on average as streams in the unburned reaches. These nutrients stimulate the growth of algae and other phytoplankton that can rapidly die and decompose, releasing carbon dioxide and methane. At the same time, the loss of these nutrients can make soil less fertile, hindering the growth of carbon-sinking vegetation.

This effect continues for at least 45 years, according to measurements by Laura Jardine, another 2017

OPPOSITE: *Green vegetation along a stream contrasts starkly with the black of recently burned tundra in the Y-K Delta.* **RIGHT:** *Rhys MacArthur collects a sample of tundra soil from the Y-K Delta to measure soil carbon and nutrients.*

Polaris student researcher. She found that organic nitrogen that had been locked up in permafrost thawed by a 1972 fire was still being converted to soluble, bioavailable ammonia and nitrate at a high rate in 2017.

Darcy Peter, a Polaris participant in 2017 and 2018, found further evidence of long-lasting warming and thawing caused by fire. Two years after the 2015 Y-K Delta fires, lowland ponds and channel fens in burned areas emitted substantially more methane on average than those in areas the fires had spared. Their water was also warmer than that in untouched ponds, and thawed soils extended deeper along ponds in burned areas than along those in unburned areas. These patterns, if they persist, suggest yet another worrisome feedback loop.

As on water, so on land. Aiyu Zheng, a 2018 participant, compared the thermodynamics of burned and unburned tundra soils and found differences there as well. Burned soils get saturated with water faster when it rains, which makes them more conductive, so they warm faster in response to air temperature. And they're also more prone to retain heat because they've lost the plants that would wick it away.

Fire rages quickly over the tundra, but its effects persist for years.

A vast maze of shallow, boggy ponds fills the floodplain along a tributary of the Kolyma River.

By sampling the chemistry of water throughout a river system, students like Erin Seybold can investigate the cycling of chemicals within the river and identify regions where permafrost thaw is contributing material to the streams.

CHAPTER 3

SIBERIA, HO!

FOLLOWING THAT CRUISE DOWN THE Lena River in 2003, Max Holmes continued to investigate the chemistry of Siberia's rivers, all the while nursing an idea that had been planted when young Anya Suslova stepped up to collect water samples. Siberia, with two-thirds of the Arctic's permafrost, was so vast, so critical to the climate future, and so little studied: "I wanted to get more people over there!" he says. And he wanted to interest more young people

Amphipods, snails, and other aquatic organisms collected from a Siberian lake

in climate science, to help build the research corps that would carry on this scientific quest after old warriors like him are gone.

That need is more urgent now than ever, says Paul Mann. He sees concern about and understanding of climate change waning in his native Britain, as polls showed they were in the States, at least before the severe weather and deadly wildfires of 2018. "I'm really frightened at how little first-year undergrads in the UK know," he says. "They don't know the difference between the ozone layer and climate cycles. And they're very swayed by the press," which elevates climate denialism and contrived dissent in the name of "fairness" and the pursuit of controversy.

The Woods Hole Research Center was established as a research, not an educational, institution. But Holmes saw an opportunity to use its resources to fulfill a vital educational mission: to interest undergraduates in climate research, jump-starting both scientific careers and future discoveries. As it happened, an old friend from Holmes's school days in Arizona proved the perfect partner for that effort. John Schade had also grown up sloshing around in the streams of central Michigan, but the two did not meet until both were in grad school at Arizona State in the 1990s. Schade set out to work in evolutionary biology, but Holmes got him interested in stream science. Since then, as a professor at St. Olaf College in Minnesota, he had developed a teaching—or antiteaching—method uniquely suited to Holmes's goals: a Socratic approach to scientific research, encouraging students to ask their own questions about how things work and then gently guiding them in designing experiments to find the answers.

St. Olaf helped seed that approach. At the college, Schade explains, "we were less prescriptive than most people are with undergraduates." The environmental studies program in which he taught was an interdisciplinary one,

LEFT, TOP: *Polaris students unload their luggage at the start of their research experience in Cherskiy.*
LEFT, BOTTOM: *Homero Peña examines a sample of vegetation extracted from a tundra study site.*
OPPOSITE: *Kelsey Dowdy works late into the night at one of the labs at the Northeast Science Station in Cherskiy.*

OPPOSITE, CLOCKWISE FROM TOP LEFT: *Jessica Eason takes a water sample from a Siberian pond; Polaris faculty member Valentin Spektor digs down to the permafrost; Erika Ramos identifies tundra plants at her study site; Andy Bunn, Polaris faculty member, shows students how to take a core sample from a larch tree.* **RIGHT:** *Scientists take water and permafrost samples at Duvannyi Yar, Siberia.*

with humanities and social-science components. Schade came to see the value of that sort of cross-fertilization and of giving students the leeway to pursue what interested them. "The program showed me if you let students take control of their education they get more done, learn more, and do better work. They feel more capable. A lot of them end up on research tracks that are more innovative than they thought was possible before."

It was a student named Sarah Ludwig, now lab manager and general factotum for the Polaris Project, who lit the spark. One day she came to see Schade, but he was too busy to talk. Instead he gave her a spur-of-the-moment assignment: go out to the wetlands around the campus, see what you see, and come back with your observations and some hypotheses to explain them.

"She came back with some hypotheses that were really quite strong," recalls Schade. "She looked at an invasive plant, reed canary grass, that was spreading along the edge of the wetlands where there had been little vegetation before, and hypothesized that this plant might be facilitating the emission of methane, spurring the microbes that produce methane in the soil by adding organic matter. She and another student then designed an experiment to test these hypotheses and found support for them. I was impressed at the sophistication of her observations and ideas and realized that these students could see the landscape in new and innovative ways, see patterns that I would miss because they were more open to new ideas, not focused by previous experience on a narrower view of the world.

"Several incidents like that showed me we need to get out of the way a little bit. We tend to give undergraduates more guidance than they need—teaching bodies of knowledge and skills, rather than process and intellectual development. . . . We need to let students dig around, let their interests drive what they seek. A big

BLAIZE DENFELD

"The Polaris project helped turn my passion for nature into a career," says Blaize Denfeld. "It gave me the confidence to pursue a PhD, helped me identify my scientific interests, and opened my eyes to different cultures and the importance of studying the Arctic."

Denfeld hadn't had any special interest in the region; it was a long way from New England, where she grew up and attended Clark University in Massachusetts, and from the earth science major she was considering. But she took a course on the region—a prerequisite at that time for the Polaris Project—was intrigued, and so joined the 2010 expedition to Cherskiy, the first of two summers she spent with Polaris. "It was good to have background information about the Arctic. But you can't prepare yourself for the experience. One of the first days there we went to a lake. One of the PIs [principal investigators] put a stick in the lake, and you could actually see the bubbles of methane coming up. This lake was full of methane. I wasn't prepared for the sight of that."

She had a background in geographic information systems (GIS) and remote sensing going in, but she saw how important it was to get a ground-, or water-, level view. "It's nice to be able to observe Earth from satellite imagery, but it is important to validate and understand how the different systems—water, land, atmosphere—connect." She decided to investigate a connection with important climate implications: the evasion (escape) of methane and carbon dioxide from rivers in the Kolyma basin.

In Polaris, Denfeld also came to appreciate the importance of diversity in science, from the wide-ranging research backgrounds of the PIs to the perspectives of the Russians she met at the Northeast Science Station, in the field, and in Polaris itself. "One particular encounter brought it all home. For my research I was taking many

"One of the PIs put a stick in the lake, and you could actually see the bubbles of methane coming up. This lake was full of methane. I wasn't prepared for the sight of that."

boat trips. Another student, Nikita [Zimov, Sergey Zimov's son and successor], and I stopped at this really remote fishermen's hut. Nikita explained that the area where they were living was getting smaller and smaller as the permafrost thawed. Right there on the river."

"Diversity makes us not only better scientists but better people. Climate change is a problem without borders, and research should be collaborative and inclusive."

Those experiences gave her "the confidence that I could work abroad." Denfeld is now completing her doctoral studies at Uppsala University in Sweden, studying the evasion of carbon dioxide and methane as ice cover melts from Arctic lakes; warmer winters mean thinner ice and more and more escaping greenhouse gases. She appreciates the international exposure: "Diversity makes us not only better scientists but better people. Climate change is a problem without borders, and research should be collaborative and inclusive."

And visible. "Science communication is more important now than ever. People in the US still don't understand the standard of scientific proof." And so she spoke at two high schools after returning from Siberia and has decided to work initially in the public sphere when she returns from Sweden, at NASA's Earth Sciences Division, coordinating research rather than conducting it.

"It's been really great to see how progressive Sweden is in terms of climate change, even in policy—they're considering a tax on meat to encourage more sustainable diets. I'd like to take this experience and apply it in the United States."

part of being a scientist is being intuitive. We want them to experience things holistically, to use all their senses."

WHRC launched Polaris in 2008 in partnership with several colleges: St. Olaf in Minnesota, Western Washington University in Bellingham, the University of Nevada at Reno, Clark University and Holy Cross College in Worcester, Massachusetts, and Siberia's Yakutsk State University, each of which provided a couple of students each summer for several years. Designated courses at those schools were supposed to prepare students for their Arctic expeditions; those who had taken them would be eligible to apply to Polaris. The faculty teaching them would accompany the students to the Arctic and then continue to work with them on their projects in the fall.

The timing was fortunate. In 2007 the International Council for Science and the World Meteorological Organization had declared an International Polar Year, reflecting the growing recognition of the polar regions' critical importance to global climate, and the US government was pouring new money into Arctic research. The next year, with funding from the National Science Foundation, Polaris brought its first crew of seven students and seven scientist-advisors to the Northeast Science Station, at Cherskiy in northeast Siberia, for four weeks of intense field- and lab work. One of two students from Yakutsk State University was Anya Suslova, following the path that began with that chance meeting five years earlier.

Holmes still marvels that they got there at all: "It's kind of crazy that we ever got funded in the first place, to take a bunch of people to the Russian Arctic. I think it was because we had a track record of working there," while so few other outsiders did.

The Pleistocene Solution

When they arrived, this first Polaris team found a pair of extraordinary hosts waiting to receive them: a

OPPOSITE: *Sergey Zimov, founder of the Northeast Science Station*
RIGHT, TOP: *Sergey and Nikita Zimov dug this cave into the permafrost to serve as a natural freezer for scientific samples as well as for their food.*
RIGHT, BOTTOM: *Nikita Zimov ferries students to the riverbank exposure of Duvannyi Yar on the Kolyma River.*

geophysicist-turned-geoengineer and radical rewilder named Sergey Zimov and his son and successor, Nikita. Sergey Zimov is the founder and director of the Russian Academy of Science's Northeast Science Station, one of the world's largest Arctic research stations. This facility has been the base for scores of outside researchers, including Max Holmes, Sue Natali, and, for its first seven years, the student scientists of the Polaris Project. Zimov is an iconoclastic and irrepressible character and an internationally renowned earth scientist whose wild-sounding ideas have a way of turning out right. "I've spent many a night in deep discussion—over vodka, naturally—with Sergey," recalls Paul Mann. "He would talk about the importance of science in building relationships between countries, of how it enables people from around the world to work together regardless of political divisions." Then, puffing on a cigarette, Sergey would argue that "because humans evolved around campfires, they need a little smoke in the lungs."

Sergey's more serious meditations on Earth and evolution have taken concrete form in an epoch-crunching, landscape-scale experiment on an expanse of grassland, tundra, and larch forest near the Kolyma River, 260 kilometers (160 miles) from the Arctic Ocean. This is Pleistocene Park, dedicated to nothing less than restoring what was once the world's most extensive terrestrial ecosystem, saving the permafrost, and protecting wild habitats and human civilization alike from a fatal collision with a changing climate.

Back in the 1980s, Sergey probed the Pleistocene chronicle embedded in the frozen yedoma. Long, tangled grass roots and stems attested to fertile steppes that gave way to modern moss and taiga. Bones, teeth, hooves, and tusks revealed a megafaunal extravaganza more like the African Serengeti than today's meager northern biodiversity. For more than a million years,

woolly mammoths and rhinoceroses, horses, yaks, musk oxen, bison, saiga antelope, camels, and deer grazed the rich grasses of what's called the "mammoth steppe." Wolves, bears, and cave lions hunted them. All that ended about 10,000 years ago, at the dawn of the warmer epoch known as the Holocene. Today, much thinner populations of reindeer graze on the tundra's lichens, and moose browse on the willows.

For decades scientists blamed a drastically cooling climate for the extinction of most of the Pleistocene megafauna, not just across Eurasia but in the Americas. But this "overchill" explanation butts up against several inconvenient facts: These extinctions happened tens of thousands of years apart on the various continents and just a few hundred or thousand years ago on newly settled islands such as Madagascar and New Zealand. Creatures such as mammoths that had survived crushing ice ages perished rapidly in the warming Holocene. Each wave of extinction followed the arrival of *Homo sapiens*. And so a contrary "overkill" hypothesis has gained traction: climate likely contributed to the demise of these giants, but it was human hunters armed with novel technologies and cooperative strategies that struck the fatal blows.

Sergey argues that this slaughter transformed not only the fauna but the mammoth steppe itself. The great grazers had lived in symbiosis with grasslands, the youngest and most productive terrestrial ecosystem. Fast-growing grasses are wonderfully efficient at turning water, nitrogen, and carbon dioxide into nutritious proteins and sugars. They can support more faunal biomass than tropical or temperate forests. Their deep roots also absorb water and anchor soils, preventing runoff and erosion. On the mammoth steppe, as on today's savannahs and prairies, the big herbivores would trample the mosses and saplings and girdle or uproot grown trees,

LEFT: *Surrounded by swarms of mosquitoes, University of Alaska Fairbanks professor and Polaris collaborator Alexander Kholodov extracts a core from the permafrost.*

OPPOSITE, CLOCKWISE FROM LEFT: *Sergey Zimov waits for a break in the weather; Sue Natali makes friends at Pleistocene Park; Capturing methane bubbles from a Siberian lake for analysis*

Sturdy Yakutian horses help shape the new tundra steppe ecosystem at Pleistocene Park.

The rolling, treeless tundra, featureless from a distance, contains a myriad of plant species.

MEGAN BEHNKE

Though she grew up in Alaska and had visited the Arctic, fieldwork on the tundra was something new for Megan Behnke, a 2014–2015 Polaris student from Juneau. "I was used to mountains protecting me," she said. "I was used to mountains on three sides and ocean on the other and being very tucked into the land. And out there there's nowhere to tuck into. But there's something about the starkness visually of big river and sky and hills and nothing to shade it. There's a lot of detail in the ground and the plants, but there's not a lot of fuzzy edges. It's very stark in a very beautiful way."

It was a fascination with chemistry, not the Arctic, that initially brought her to Polaris and Siberia. "I was a restless sophomore chemistry student [at St. Olaf College in Minnesota] when I stumbled upon the Polaris Project," she explains. "The theories and mechanisms of organic chemistry enchanted me, but I hated the structure of the lab, twenty-five sophomore students with Bunsen burners trying to make a molecule by following a recipe. I went hunting for a way to pair what I loved about chemistry with something more satisfying and wound up in the office of John Schade, who told me about biogeochemistry and suggested I apply to go to Siberia with him and the Polaris Project."

Behnke reveled in the adventure, "living on a barge with a janky sauna, a common room, some bunk rooms, and a lot of mammoth bones, parked on a gravel bar in the Kolyma River." And stream biogeochemistry—which married her passions for chemistry and backcountry exploration and introduced her to the interdisciplinary science of ecology—"was a revelation." She probed the photochemistry of dissolved organic matter released from thawing permafrost—how it's transformed through exposure to light—and has continued to track it in peatlands, glacial streams, and temperate rainforests as well, through her current PhD studies at Florida State University. The same questions still possess her: "How do organic molecules change as they move through a system? And what does that mean for the ecosystem? How good do they taste to microbes and why?"

That was one world the Polaris Project opened for her; the other was the wide world itself, "experiencing international collaboration for the first time," as she puts it. "Living and working with Russian scientists, and the German, Chinese, and Scandinavian scientists who were at the [Cherskiy] station at the same time, was formative. My research mentality was formed with input from many nations and work styles."

Behnke has tried to pass on what she learned in Siberia, speaking to high school students, to nonscientists on a weeklong Global Rivers Observatory cruise up the Columbia River, and to whoever asks her what she does. "When climate change or my research comes up, I usually start by calling myself an 'adventure chemist,' then move on to describing the crazy places I work—the giant cracks running through large cement buildings, houses crumbling, drunken trees tipping over from thermokarst activity, and most of all mammoth tusks poking out of slumping riverbanks. Then I can explain why these things are happening—the permafrost is thawing, climate change, etc. But the images themselves are initially engaging and not polarizing the way the phrase 'climate change' can be."

The rapidly shrinking glaciers outside Juneau are even more visible testimony to the change underway. "Living in Juneau, I grumble about all the tourists, but people should *see* glaciers, appreciate them. See these big changing ecosystems." And the smaller systems closer to home. "Anywhere there are going to be things shifting because of climate. Being aware of ecosystem changes is the first step.

"I don't know how much we can actually change what's happening to these ecosystems right now, but I do know if we don't do everything we can, it will get so much worse. We can prevent that. There's no excuse not to do everything we possibly can."

"I don't know how much we can actually change what's happening to these ecosystems right now, but I do know if we don't do everything we can, it will get so much worse."

making space for resilient, fast-growing grasses. Their wastes would fertilize more grass growth—a nutritional feedback loop.

And thus to Pleistocene Park's bold mission: To show that it's possible to recover the mammoth steppe by removing mosses and trees so native grasses can grow and reintroducing the grazers that sustain the system. And then to extend that success with a network of restored tracts across Siberia and North America, until they join to reconstitute a Beringia-wide steppe protecting the permafrost and stabilizing the climate.

The effort is still at a formative stage: Sergey and Nikita Zimov have introduced shaggy wild Yakutian horses, European bison, yaks, sheep, and musk oxen (the last captured after an iceberg-dodging voyage to Wrangel Island and a polar bear–dodging chase across its foggy wastes). Reindeer and moose already inhabit the surrounding landscape. But much remains to be done. The current density—100 to 120 big herbivores on the 20 square kilometers (8 square miles) that have so far been fenced of the 144 square kilometers (56 square miles) that comprise Pleistocene Park—still falls well below the goal of 20 per square kilometer, and even farther below the densities found on Africa's richest grasslands. Predators—wolves, bears, Amur tigers, perhaps cold-adapted cougars from Canada—will be needed to manage the grazers. The predators induce the herbivores to clump into herds and keep moving and spreading their manure, rather than staying put and overgrazing and overfertilizing the same pastures.

How to Make a Mammoth

One keystone species will still be absent from the neo-Pleistocene landscape, however: the apex herbivore that gave the mammoth steppe its name. Like today's savannah elephants, woolly mammoths were

the supreme gardeners and geoengineers of their grassy domains. Elephants can smash and uproot trees at a furious rate as they scratch their backs, seek out tender leaves and inner bark, and vent hormone-driven aggression. One study found that elephants in Tanzania's Serengeti wrecked an average of three trees every four days. Managers initially blamed the desertification of much of the Serengeti in the 1970s on the elephants' destructive ways, but closer study revealed the actual culprit: overeager burning by human herders trying to improve the pasture. After the burning was controlled and the Serengeti's grazers—wildebeest especially but also elephants—rebounded from disease and poaching, the grassland recovered.

In Siberia, the mammoths' presence can still seem tantalizingly near. Thousands of tusks and uncounted bones have emerged from thawing permafrost and tumbled out of crumbling riverbanks; the tusks support a lively, legal, sometimes violent trade in ancient ivory. Mixed with all the other Pleistocene detritus are bits of skin and hair from ancient elephants. Occasionally miners or builders dig up a near-intact mammoth carcass, and dubious tales persist of czars and explorers feasting on freezer-burned 30,000-year-old steaks.

These discoveries have fueled a dream that began as a spoof in MIT's *Technology Review* in 1984 and gained respectability as cloning and gene-editing technologies advanced: to resurrect or recreate the woolly mammoth. At first, proponents imagined cloning an actual mammoth, à la Dolly the sheep, or inserting its DNA into an egg from a living Asian elephant. (The two species are closer cousins than today's Asian and African elephants.) But these hopes crashed against Methuselahn realities: however well it's been frozen, ancient DNA emerges in bad shape after millennia-long battering by cosmic rays.

LEFT, TOP: *Sergey Zimov found this jawbone, likely from a reindeer, on the bank of the Kolyma River.*
LEFT, BOTTOM: *A baby mammoth replica greets visitors at the museum in Yakutsk, Siberia.*
OPPOSITE: *Polaris students collected these Pleistocene remains in a single day of exploring at Duvannyi Yar—a megafaunal extravaganza, like Africa's Serengeti.*

Today, a Harvard geneticist named George Church is taking a less purist but more plausible approach: tinker with the Asian elephant genome, inserting genes for Arctic-friendly traits such as cold-resistant hemoglobin, small ears (floppy elephant ears function as heat-releasing radiators), thick body fat, and even thicker fur. In 2017 Church said he hoped to deliver a quasi-mammoth to Pleistocene Park in a decade. He even looks toward mass-producing mammoths in the lab, since elephants breed too slowly to provide the shock troops needed to stop the permafrost's thawing.

The would-be mammoth makers have yet to answer some very big, perhaps irresolvable questions. Altering genes, with all the ethical and prudential questions attending on it, is one thing; how do you reproduce the rich family bonds, cultural legacies, and imparted knowledge that define elephant life as much as tusks and trunks do? Other questions loom regarding the whole Pleistocene Park concept. How much would all the methane emitted by rampant megaherbivores cut into the climate-stabilizing benefits of restored mammoth steppes? No one, not even the Zimovs, has rigorously calculated the net benefits of restoration, but Nikita is confident they would be substantial. He notes that the 90 to 110 million tonnes of methane Pleistocene grazers are estimated to have belched each year pale against the nearly 400 million tonnes emitted by wetlands in our warmer, wetter Anthropocene climate.

A Constellation of Grasslands

Pleistocene Park is not a total outlier. It's one among many efforts to restore grassland ecosystems across the world, from the American Midwest, Southwest, Northwest, and Eastern Seaboard to Australia, central Europe, central Asia, South Africa's high veldt, and the South American pampas. Generally, however, these restorations are directed primarily to recovering wildlands and habitats, not to preventing carbon releases, though they may also have that benefit. Often they are undertaken on land that has been exhausted by agriculture or, as on the dry, cold, rapidly depopulating northern plains, was marginal farmland to begin with. There especially, native herbivores, in particular the bison that once dominated the prairies, are seen as critical to restored ecosystems. Paul Martin, the paleobiologist who first proposed the overkill hypothesis, and biologist David Burney went further in their 2000 essay "Bring Back the Elephants." They proposed "the ultimate rewilding": releasing Asian and African elephants in warmer parts of the American West and on South America's savannahs—a beguiling notion, but one that would probably require massive dedications of wildland and even more forbearance than that required to live with wild wolves.

Other American conservationists envision regional restorations on a scale echoing Sergey Zimov's vision of a constellation of Pleistocene parks across Eurasia and North America. The Yellowstone to Yukon Conservation Initiative works to knit together and safeguard a continuous wildlife corridor along about 3,200 kilometers (2,000 miles) of the US and Canadian Rockies; such corridors can help vulnerable species adapt to climate change, enabling them to move uphill or northward as conditions warm. The Buffalo Commons, proposed in 1987 by the academics Frank and Deborah Popper, would dedicate 360,000 square kilometers (139,000 square miles) of semiarid plains across ten states, from Texas and New Mexico to Montana and North Dakota, to "more bison and less cattle, more preservation and ecotourism and less conventional rural development and extraction."

OPPOSITE: *In the Pleistocene era, the Siberian Arctic was a tundra steppe ecosystem, the world's largest land biome at the time. Even today, grasses such as this foxtail barley* (Hordeum jubatum) *can support more animal biomass than tropical forest.*

OPPOSITE: *A European bison surveys its new home at Pleistocene Park.*
RIGHT, TOP: *Polaris student Peter Han takes notes on his observations at Pleistocene Park's grasslands.*
RIGHT, BOTTOM: *A young moose, raised in captivity, forages at the entrance of Pleistocene Park.*

As the Poppers have since noted, this is an idea with many historical antecedents, going back to the Great Plains Park proposed by the frontier artist George Caitlin in 1842. Their vision is still far from full realization, but it continues to inspire public and private initiatives across the West. One self-acknowledged heir to it, the Great Plains Restoration Council, has completed or launched prairie projects in Texas, New Mexico, and South Dakota and an innovative "Restoration, not Incarceration" project, recovering both ecosystems and young lives caught up in the legal system.

But Nikita Zimov's hopes of extending the Pleistocene Park approach to North America have dimmed as he's seen more of how things do or don't get done in this country. He and his colleagues are currently fighting to get bison flown from Alaska to Pleistocene Park. "It is a disaster to deal with US air companies," he says, an opinion many American airline passengers might second. "Dealing with that matter I realized that it would be impossible to make a park similar to ours in the US. Too much bureaucracy."

What restorations have been undertaken lie in temperate-zone grasslands and, in the case of the Yellowstone–Yukon corridor, forests. They would have unquestioned benefits: better habitat for endangered species and other fauna, reduced flooding, less groundwater loss and soil erosion, more carbon storage as deep-rooted perennial grasses and forbs replace shallow-rooted annual crops. But only Pleistocene Park proposes to safeguard the critical carbon bank embedded in the Arctic's soils. And it sets out to do that on the front line of the emerging crisis: Russia, where most of the permafrost lies. "In Alaska," Sergey Zimov said after visiting the largest Arctic research station at Toolik, Alaska, "scientists fight over kilograms of carbon. In Siberia, we have gigatons per scientist."

Owing to the short growing season, Arctic trees have very narrow growth rings. The oldest tree section collected by Polaris scientists near Cherskiy was about three hundred years old but the trunk was no more than a foot in diameter.

A slurry of ice bits lies outside a core hole drilled into the permafrost.

Polaris students hike along the bank of the Kolyma River, Siberia.

Nickolai Torgovkin examines a permafrost sample in one of the Northeast Science Station's labs.

Scientists have excavated permafrost tunnels for research under the Melnikov Permafrost Institute in Yakutsk, Siberia. The Polaris Project sometimes visited this site while traveling to Cherskiy.

Ice crystals grow on the ceiling of the Yakutsk permafrost tunnels.

CHAPTER 4

IN ALASKA AND ON THIN ICE

SINCE THOSE FIRST YEARS AT Cherskiy, the National Science Foundation (NSF) has funded the Polaris Project twice more for three- to four-year rounds of summer research expeditions—a nearly unheard-of success in the competitive scramble for NSF grants. That reflects in part the tangible research results the project has achieved by letting students take the lead.

Polaris student and Alaska Native Darcy Peter collects soil samples in the Y-K Delta.

John Schade drives home the point: "Undergraduates in the Polaris program aren't obliged to perform 'successful' experiments. Not all good ideas pan out or end up as publishable products, but that's okay. We learn as much or more from projects that fail in the end, and the intellectual impact on the students is still strong because it builds from the development of the good idea, not the publication of it.

"Graduate students and postdocs have to worry about productivity, and rightly so," adds Schade. They must stay focused on narrow, defined goals, on getting the data they need for their dissertations. Undergrads, by contrast, "come in with open minds," says Sue Natali. "They can take risks, explore the landscape, build their hypotheses"—ask questions whose solvability may be uncertain, but whose answers, if found, may be all the more valuable as a result. "I'm always learning something from their research," she says. That means knowing when to hold back and give them a long lead, rather than taking charge: "I pretty much serve as a glorified field assistant and help them do *their* projects."

The funny thing, says John Schade, is that by taking chances and letting the students lead, "we get a better science outcome; we accelerate our understanding of the Arctic." Polaris undergrads have racked up an impressive tally of research publications and conference presentations. As of 2018, they had published 11 papers in scientific journals, made 79 presentations at major scientific conferences, and coauthored 14 more papers and 17 more presentations. A case in point: Nigel Golden, a student from the University of Massachusetts who joined the 2014 mission to Siberia, was interested in wildlife, a field that had not figured in the Polaris Project's planning, nor in its resident scientists' thinking. He'd hoped to study caribou, but there were none at the Cherskiy research station. There were, however, many ground squirrels.

What did ground squirrels have to do with permafrost and the carbon cycle? Golden set out to find out. What he found rocked not just scientific circles but international media: the squirrels' extensive burrowing exposed the frozen soils to warm air in summer months, while snow cover blocked cold air from them in winter. The result: faster thawing. The squirrels' feces and urine fed the growth of more microbes, which produced more carbon dioxide and methane.

After Golden presented his findings to the American Geophysical Union, the BBC picked up the story and other media followed. Now "rodentogenic" carbon emissions have joined the factors climate scientists consider in modeling future changes, and Nigel Golden is completing his doctoral thesis on them. "I *never* would have thought of that," says Natali. "I know wildlife are there, we see them, but we never would have done that without Nigel."

During the 2018 Polaris session another student took a flyer, asking a question about something her seasoned elders took for granted, and scored a breakthrough that surprised her as much as anyone else. Alexandra Lehman wanted to determine whether the biological character of small tundra ponds varied as sunlight levels changed through the day. This entailed trekking out to three different ponds every three hours across two full days to take water and gas samples. "I wish I could have done one more day," Ali Lehman says, "but it's very strenuous staying up for twenty-four hours," even in the endless day of a boreal summer night. She already felt "blinded by the sensory overload" and overwhelmed by the rigors of conducting accelerated field research in a place so different from her native Baltimore and beloved Chesapeake Bay—"*completely* different."

OPPOSITE, CLOCKWISE FROM TOP: *Arctic ground squirrels near Cherskiy; Ali Lehman labels a sample marker on Alaska's Y-K Delta, on the way to her own unexpected breakthrough; Polaris student Nigel Golden collects soil outside a ground squirrel burrow near Cherskiy.*

LEFT, TOP: *Kenzie Kuhn collects a water sample from a small pond in the Y-K Delta.*

LEFT, BOTTOM: *Scientists examine an active layer soil sample on the Y-K Delta.*

For two weeks after the student researchers return from the tundra, WHRC's ground-floor auditorium becomes a scientific boiler room. There, lined along long banks of tables, they labor over their laptops, samples, and GIS maps while the senior scientists stand ready to advise and encourage, and prod them with questions when they get stuck. When Ali Lehman analyzed her samples and assembled her data, her stress turned to despair. The biological factors she'd set out to measure—algae, chlorophyll, oxygen, and nutrient levels—didn't vary significantly across the day: "All of that just did not show anything. So yeah, that was kind of a failure." Then, with the deadline to report her research looming, she measured the methane in her water and gas samples and was relieved, almost exultant. Methane levels declined steadily through the day, in concert with the amount of ultraviolet light the water absorbed, reflecting the amount of sunlight it received.

"This is *really* interesting data," John Schade said when he saw Lehman's methane findings. "Nature just handed you an experiment!" Afterward he explained: "She found significant variation in methane flux across the course of the day. I didn't think that would be true—it was a limitation on my part. We thought we only needed to sample [sites like this] once a day."

It got better. Lehman's findings suggested not only that sunlight levels affect methane emissions but why they do. Ultraviolet absorption levels indicate the complexity of the organic carbon rings dissolved in the water. The microbes that produce methane feed more readily on shorter, simpler carbon rings. More ultraviolet exposure means shorter carbon chains means more microbial activity means more methane. These findings comport with those of another 2018 Polaris participant, Joshua Reyes. In the lab, he subjected sterilized samples of tundra water to varying levels of artificial sunlight and found that even without microbes present, the light "photomineralized" organic carbon compounds, a precursor to the release of carbon dioxide.

Wind provided an exception that seemed to prove the point; it blew on one of the days Lehman sampled, altering the pattern of methane emissions. As it happened, the Polaris team had installed a weather station nearby, which provided simultaneous wind readings to correlate with the methane measurements. "Putting all these data together, there was really a strong story," says Schade. "This is a good example of how observation leads to new questions." Heady stuff for an undergraduate's "failed" field research.

Moving On

Even as they sweat over their own data and reports, the students pause to listen to and advise on each other's travails, cheering findings, lamenting pitfalls, and suggesting solutions. That collaboration grows even more intense in the field, where not just research success but life and safety can depend on it, and members must travel together in the treacherous terrain.

This collaboration came especially easy to Natalie Baillargeon and Rhys MacArthur, fellow students at Hampshire College who joined the 2018 Polaris expedition together with their faculty advisor, Seeta Sistla. Each assisting the other, they undertook complementary projects, investigating plant recovery and soil characteristics following tundra fire.

This kind of transition from classroom to Polaris was the norm in the project's early years. The fact that participants were funneled through a few partner colleges, with instructor-advisors from their home schools joining them in Siberia, greased the wheels of cooperation. But this homogeneity was also limiting; common experience meant limited experience. If Polaris students were to be proxies for their peers, on the front line of the next generation's confrontation with climate change, they should be more representative of that generation. But the students applying, however able, tended to come from similar upper-middle-class backgrounds.

Furthermore, integrating Polaris projects and campus coursework had not paid off as Holmes and Schade hoped. Despite all their prior preparation, the students and their advisors did not get down to the nuts and bolts of developing their projects until they actually arrived at the Siberian station.

And so in 2011 the Polaris Project shed its college partnerships and developed a new modality. Instead of coursework, a series of online meetings from March to June and an intensive weekend at Woods Hole in April now prepare participants for their summer expedition. As preparation for the first online session, Schade primes their experiential pumps by asking each student to do as he had asked Sarah Ludwig to do back at St. Olaf: "Go to an ecologically significant place, make some observations, and bring them back to share with everyone." Use all your senses. Stop and see what's in front of you. Ask why it is the way it is, he urges them.

For subsequent sessions, the students read a heavy load of research papers on permafrost, carbon exchange, and related issues—initially assigned, then selected from a menu of offerings. "We have to give them scientific boundaries," says Natali, "and these are basically set by the papers we assign them." At the meetings, Schade, the master intellectual emcee, teases out the insights and interests suggested by the papers. Did anything in the methodology or findings capture their imagination? What questions did it raise, what directions did it suggest for their own projects? As other students chime in with suggestions and observations, each one's turn to

INTEX

OPPOSITE, CLOCKWISE FROM TOP LEFT: *Mia Arvizu shares results with faculty member John Schade; Maggie Powell collects a water sample in the driving rain; the Polaris team floats science gear across "Landing Lake," as the team calls it; Natalie Baillargeon sorts a tundra plant sample.* **RIGHT:** *The Y-K Delta is a maze of ponds, lakes, and river channels.*

respond becomes both a spotlight performance and a contribution to a collaborative inquiry.

At the same time, the Polaris Project opened the door to undergraduate applicants from anywhere. Strategic networking got the word out to groups underrepresented in science, attracting a rich infusion of diversity. Half or more of the participants in recent summer expeditions have been African American, Native American, Alaska Native, or Hispanic students—including several from hurricane-battered Puerto Rico. For them, climate change is neither an abstract nor a distant threat.

Working at the Northeast Science Station, in the middle of the biggest permafrost expanse of all, offered many advantages: a close-up view of the thrilling Pleistocene Park experiment, friendships and fruitful exchanges with Russian students and scientists, the deep knowledge that the Zimovs and their colleagues had developed in more than 30 years working at Cherskiy, and the convenient infrastructure—from roofs to showers to Wi-Fi—of an established research station.

But in 2015, when the Polaris Project's second four-year NSF grant drew to a close, that door slammed shut. With war raging in Ukraine and the United States decrying Russian meddling there, tensions between the two nations were rising fast. This did not exclude the project from Siberia directly; American scientists were still welcome. But the National Science Foundation seemed apprehensive about continuing to fund research in Russia. Though Holmes's original goal had been to build scientific bridges to Russia, the Polaris Project would have to leave there and return to America.

Landing Lake

Alaska's North Slope was the obvious destination. The climate there most closely resembles that of northeast Siberia, and the Arctic's largest research station, with

OPPOSITE: *The Polaris field camp in Alaska*

berths for 100-plus researchers, is located there at Toolik. For that very reason, however, this is the most heavily studied part of the Arctic. "At a well-established research station there are already so many people," says Sue Natali. "So much has already been done. But if you go to a new place, any data you gather are new."

And so the Polaris team looked to an understudied territory with special significance: the Yukon-Kuskokwim Delta, the wide alluvial plain formed where the fifth- and ninth-largest rivers in the United States meet the Bering Sea. The Y-K Delta lies at the Alaskan Arctic's southern margin, below the Arctic Circle, in the transitional band of what's called "discontinuous permafrost." This is the front line of the seesaw battle between freezing and thawing. It is here that warming's effects will show first.

Holmes and his colleagues initially sought a research site near the Native village of St. Mary's, but the permitting arrangements proved too complex. In Russia, a single central authority could grant permission to do research; here, a maze of overlapping jurisdictions and Native communities had a say, and boundaries and authorities were not always clearly defined. So the Polaris team struck out into federally managed wilderness and found a site beside a small floatplane-ready lake—currently called, appropriately, Landing Lake, though they hope to find and adopt its original Yup'ik name.

The living arrangements were rustic enough to make the Siberian science station seem like a luxury resort; forget about showering or checking your email. Everything had to be flown in by helicopter or floatplane, including the removable boardwalk laid down between camp facilities to protect the fragile tundra from trampling. Everyone shared the same bucket toilets. Some would dip their heads over the back of a rubber dinghy to get the closest thing to a shower available. But they couldn't stay submerged for long because of the leeches.

On their first full-scale expedition to the Y-K Delta, in 2017, the Polaris team took no gun or electric bear fence, just large cans of pepper spray to repel any intruders. One large grizzly came near the camp but did no damage. A bear fence was promptly flown out and installed around the galley and lab but not the sleeping tents. Each student and scientist gets trained in using bear spray before embarking. Also in fire-starting, wilderness safety, pitching a tent in the wind, and other essential expedition skills. Each must learn to endure the clouds of mosquitoes that harry them like wolves stalking caribou. "When it was breezy they got blown away, but when it was hot and still—oh god!" Rhys MacArthur recalls with a shudder. "I lived in my bug shirt, even though it was really hot and stiff. I'd never used DEET in my life, and I used it all the time. I'd never used a bug net, but I lived in that thing."

Natali finds that, far from hindering the Polaris mission, the camp's remoteness and primitive conditions actually bolster it. Cooperating to survive in the wilderness and watching each other's backs prime the students for helping each other's research. The lack of internet access eliminates not only a major source of distraction but also easy research shortcuts. No more looking it up on Wikipedia. "If you want to know what's going on in the landscape, you have to *look*," says Natali. At day's end, instead of retreating to their computers, scientists and students alike "sit around the table, exchange experiences, and generate ideas."

"The undivided attention breaks down the barriers of the academic world," says Paul Mann. "When we [senior scientists] first met them, they were very daunted. But out there everyone's equal."

KELLY TURNER

At forty-three, Paul Mann's student Kelly Turner was about twice as old as her peers on the 2018 Polaris expedition. She already had another career, two children, and a lifetime's worth of experience under her belt. After growing up in an English seaside town, she left to work as a registered nurse in Western Australia, much of the time with Aboriginal people, including a stint in the Outback. "They were still so close to the land," she recalls, "eating only what they could take from it, not farming it for their own gain—making fishnets, casting from the shore, and making amazing use of fire as a hunting aid and to control wildfires. They feel themselves to be part of the land. Their sense of custodianship sparked my interest in the environment." It also inspired her to return to school, and then to venture into the Arctic.

Turner's nursing background informs the research she's chosen to pursue in the branch of environmental health known as exposure science. She is investigating one little-studied but potentially serious effect of thawing: the release of toxic pollutants, in particular mercury, impounded in frozen soils. Mercury from coal-burning power plants and other industrial and mining operations already wafts north and gets drawn down by tundra vegetation. As frozen soils thaw, microbes convert the inorganic mercury impounded in them to highly toxic methylmercury. This washes down to the sea and other water bodies and accumulates, sometimes at extremely high levels, in Arctic fish and birds and especially in animals at the top of the food web, such as belugas, seals, polar bears, and humans.

How much more mercury will warming cause the soils to release? Where will it go? How will it affect humans, especially Alaska Natives whose diets center on fish and marine mammals? Answering these questions requires determining how the mercury stored in permafrost reaches the waters—an uncharted process. Turner set out to tackle it by sampling lakes, ponds, surface soils, and permafrost. She had to send her samples to a lab in Germany to get their mercury levels measured, and the initial results suggest a host of new questions to sort out. But one finding is clear: "I love the Arctic. I'm fascinated by the place, really interested in the indigenous communities and the impacts of climate change on them. This has given me a huge sense of direction."

Natalie Baillargeon prepares to measure the permafrost thaw depth along a transect in the Y-K Delta, Alaska.

Around them stretches a patchwork of tundra, fens, lakes, and streams very different from the surging rivers of Siberia. Land, ice, and free water are closely interwoven here. "It's quite hard in Siberia to connect what's going on, on land and in the water," Paul Mann explains. Siberia's rivers are distinct entities, cutting like highways through the dense taiga. "In the Y-K Delta it's not hard at all." Meandering streams and ponds are so closely interwoven with the broken permafrost that "you can almost 'see' how water moves through the landscape. It's a biogeochemist's dream!"

Aquatic-terrestrial connections came to the fore within the Polaris Project as well. The project's emphasis had initially been aquatic, reflecting Max Holmes's and John Schade's focus on stream chemistry. That changed with the ascension of Sue Natali, an Arctic soils specialist who started with the project as a postdoctoral researcher in 2012 and became expedition leader in 2014.

True to the Polaris precept that diverse life experiences enrich the scientific enterprise, Natali came to science by an unusual route. After finishing undergraduate studies and working for several years, she took a decade off and then, rediscovering a passion for science, returned for graduate school. At first she turned to plant science, examining how carbon dioxide levels affect the uptake of mercury and other trace metals in forests. But the connections between plant and soil processes led her to investigate the soils the plants grow in as well. Nowhere did soil science seem more relevant than in the Arctic, where the state of the permafrost might determine the fate of the world's climate. And so she came north and undertook several permafrost research projects in Siberia and across Alaska.

As a plant and soil scientist, Natali brings valuable expertise and a much-needed counterweight to the aquatic focus of the other principal scientists. She

finds the encounter equally enriching to her own work. “There’s so much aquatic and terrestrial scientists can share,” she explains. “They influence each other’s work.”

Moisture and heat are the great drivers of changes in permafrost. Parsing the combined effects of these two factors is key to understanding how a changing climate will affect the permafrost and how thawing permafrost will in turn affect the climate.

Land and water, freeze and thaw. These are the dynamic dualities that shape the tundra environment. Those who live in it and still live on its fish, game, berries, and other natural riches stand poised between both. Lately that’s become like trying to balance on a shaking teeter-totter.

LEFT, TOP: *Floatplanes are used to ferry supplies and scientists to the Alaska field site.* **LEFT, BOTTOM:** *Schoolmates and research partners Natalie Baillargeon and Rhys MacArthur watch the floatplane land, Polaris field camp.* **OPPOSITE:** *The sun rises behind caribou antlers on the Y-K Delta.*

Clogged Rivers and Sagging Houses

That balance is not an abstract concern for Darcy Peter, who comes from the village of Beaver, about 200 kilometers (130 miles) north of Fairbanks in north-central Alaska. Acutely aware of the urgency of the challenge facing her Arctic homeland, which is warming at twice the average global rate, she has been a sort of conscience to her Polaris colleagues, reminding them of the human impacts of the phenomena they probe.

“What good is research if it doesn’t benefit anybody but the researchers?” Peter asked them. “I’m from a small village in Alaska where people [from Outside] were always coming and going and we never had any idea what they were doing,” she explained to fellow members of the 2018 expedition when they met by a Skype conference call before embarking for Alaska, “and I don’t want us to be like that.” So she arranged meetings between the Polaris team and the residents of the Native village nearest their camp and of Bethel, the metropolis (by western Alaska standards) 88 kilometers (55 miles) to the southeast. “I thought we might share

DARCY PETER

From the time she was born, Darcy Peter would accompany her uncle Tuffy as he fished and hunted along the Yukon River. Their home village, Beaver, lies a thousand miles up the river from the delta where the Polaris team works. He and Auntie Ai hunted, fished, and trapped in traditional fashion and also operated a smokehouse, bed-and-breakfast, and tour company that brought much-needed income to isolated Beaver: "They were the backbone of the community."

About 18 years ago, when Peter was six or seven years old, Uncle Tuffy turned to her on the boat and said, "Darcy, the weather is getting warmer. The river is getting wider and shallower."

"I just checked out, because I thought he was crazy. I didn't understand the implications until I was in college." Years later, as a student at Fort Lewis College in Colorado, she took an ecology class. "We talked about the changes that can occur as a result of climate change. I called him up and said, 'Uncle Tuffy, you were so right!'"

Since then, those changes have become evident to all—and dire for communities like Beaver. The Yukon, a vital waterway in summer and snowmobile passage in winter, started taking until late January, even February, to freeze over. "This year, for the first time as far as I know, the river didn't freeze completely—which is horrible. We can't go hunting, can't go trapping, can't collect firewood, can't go anywhere. It impacts the village's morale and economy. It's increased our reliance on Western things like buying firewood and other fuels, which require money that people don't have." Or, when they can't gather dead wood along the river, they cut down live trees around the village.

"This year, for the first time as far as I know, the river didn't freeze completely—which is horrible. We can't go hunting, can't go trapping, can't collect firewood, can't go anywhere."

Thermal erosion—thawed soil washing into the river—clogs channels, making the river spread wide and shallow. It swallows the beaches where fishermen launched their boats and cuts steep, forbidding banks. It's become hard to find spots deep enough for fish wheels—ingenious water-driven salmon traps assembled from stripped poles. There are far fewer fish anyway.

"The science lets me give numbers and back up what my people are experiencing."

"Three summers ago I asked, 'How many fish did you get, Uncle?' He said, 'We got fourteen in the whole season.' I remember when we would have twenty or twenty-five salmon each time we checked the wheel."

These changes, observed and felt, have fired Peter's determination to pursue a scientific education and career. She knows that, like the salmon, she's swimming against the current: "Not many Alaska Natives go on to PhDs or master's degrees, and if they do it's in rural development or economics. Those are important, but I got really interested in permafrost and carbon emissions. And outside Toolik, which is way far north, there's not much research being done on them in rural Alaska." The Polaris Project is a rare exception, investigating climate effects on the particularly vulnerable discontinuous permafrost on the tundra's edge. Peter wants to bring what she's learned there home to the boreal-forest environment she grew up in. "The science lets me give numbers and back up what my people are experiencing."

She realizes that the warming climate may not be so much on people's minds elsewhere. "It's really easy to get wrapped up in the day-to-day bubble of your life. In some places not getting the right nail color can ruin a person's day. But struggle in Alaska can be 'I don't know where my next meal is going to come from.'

"What [people elsewhere] do is impacting others, but eventually it's going to impact them. This is not something that only Alaskans are going to experience. Look at all the hurricanes and floods that are happening.

"My wish would be for people to educate themselves. Understanding can change the way a person lives."

what we're finding, in terms that everyone can understand, and maybe receive some traditional knowledge from them."

At one Bethel meeting, held just before many of the students had to catch a flight back to the Lower 48 and recorded by Krysti Shallenberger of the local radio station KYUK-FM, Sue Natali explained why the Polaris Project had come to the Y-K Delta, what they were investigating, how that fit into the broad picture of global warming, and how they hoped both to share what they found and learn from residents who'd been there much longer than they had. Student Kelly Turner, who had come to environmental science and Polaris after a first career as a public-health nurse, talked about trying to combine the two disciplines by studying the health impacts of changes in the Arctic environment and diet. Finally, Paul Mann put the question directly: "What changes have you seen, and what should we do?"

"Interview the elders," one woman shot back. "Maybe bring a family member to help translate." She and other nonelders went on to recount the changes they'd seen in their shorter lives. "It really impacts people's ability to travel safely [on firm ice and snow] and gather food," one said. "You have to wait longer to go out and stop going out sooner" in spring.

"It used to stay cold for a week," one man recounted—meaning 20 to 30 below zero, sometimes even lower. "Now it's just for a day or two. I was out at fish camp and I was thinking how I used to have to clean brush out every five years. Now it's every year—everything's growing so much faster. We used to have to level our houses every five years. Now I did it last year and it needs it again." As the ground thaws and turns to goo, foundations shift and sink, rooflines sag, and homes become "smiley-face houses."

OPPOSITE: *A tidy array of soil samples awaits analysis.*

"Leveling a home is not cheap out here," another resident added. He recalled how 20 years ago if you tried to dig a hole you would soon hit frozen ground and have to wait for it to thaw. "This year I found little sinkholes in my sand pad where my whole foot could go in. There's a lot of erosion. That's kind of scary."

That erosion has transformed the rivers—vital transportation arteries—as well. It fills channels, making streams wider and shallower, harder to cross and to navigate. "It's making it harder for people to get to their berry camps and their fish camps," one woman explained, "because of all the erosion that's getting put into the water, the sandbars and everything. It's getting scary because when we get high tides we can't see the sandbars and things poking up from the sandbars. Now people talk about going around on the coastline just to get to the berry camps, which is a lot longer. It's getting spendier just to harvest berries [and] harder to get to fish camps on small rivers, with the high banks."

The fish themselves aren't what they used to be. Their skin doesn't peel off as easily, perhaps because they're less fatty, and, one man reported, they're "dying faster in the nets."

"We used to not have such a fly problem in June when we're cutting our fish to dry, but this year it's really bad," another woman lamented. "We're noticing too that the thaw and spring come earlier. Animals like rabbits are still white. They're visible to predators. We wonder how they're going to survive, because they can't adapt so easily."

Across the Arctic, it is the same story: ancient cycles of ice and thaw are unraveling before people's eyes. In Anya Suslova's Siberian homeland, plants and animals never seen before are appearing as the weather warms. The Lena River is freezing later in the year, so the ice does not thicken as it used to: "With the thinner ice, fishing is really dangerous," she says. Even the seasons are changing. "It used to be just winter and summer. Now we have these shoulder seasons. People want to know what future predictions are, but there's no information."

The Polaris Yukon-Kuskokwim Delta field camp occupies a small peninsula jutting into a shallow lake.

A Savannah sparrow brings insects to its young, concealed in the tall grass near a lake in the Y-K Delta.

CHAPTER 5

A NEW GEOGRAPHY OF HOPE

PERHAPS NOTHING DEMONSTRATES the interconnectedness of Earth's natural systems so urgently as the carbon time bomb ticking under the tundra and taiga. Permafrost is a phenomenon that few people have seen, much less studied, and that few give any thought to. Many dwellers at lower latitudes have never even heard of it, and if they did, they would never imagine that it could

Wildflowers bloom on a cliff top along the Kolyma River, Siberia.

affect their lives. And yet, as it thaws, the Arctic's permafrost has the potential to upend the lives of people living in seaside condos in Miami, in exurban dream houses overlooking scenic wildlands in California, on the scorching and artificially green once-and-future deserts of the Southwest, on shrinking coral islands in the Indian and Pacific Oceans, and in flimsy houses perched precariously on slippery hillsides in Haiti and on the floodplains of Bangladesh.

Vital though they are to biodiversity, coral reefs and tropical rainforests store far less carbon than the frozen ground. But they are visible and spectacularly beautiful—the landscape equivalents of charismatic megaspecies. The tundra is suffused with a quieter yet no less haunting beauty, but the ground beneath it is by definition out of sight. Only when it thaws and erodes at the edges—muddy cliffs collapsing onto beaches and riverbeds—does permafrost reveal the treasures locked within it, from ancient pollen grains to Neanderthal artifacts and mammoth tusks. It is a story that takes more telling, and the time for telling grows ever shorter.

And so the senior and student scientists in the Polaris Project and their counterparts in the rest of the tiny permafrost research community scramble to unravel that story and find ways to present it to a wide, distracted audience allergic to more bad news. All the while they carry the knowledge of the dire consequences ahead if we do not reverse the trajectory of our carbon emissions and the feedback loop of "warm, thaw, release, more warm, more thaw, more release" sets in—and of all the obstacles to concerted action embedded in our politics and economics, and in human nature itself.

How do we face such facts without throwing up our hands and taking refuge in despair? Once, Sue Natali, accompanied by Max Holmes and another scientist, briefed a group of US State Department officials on

LEFT, TOP: *Megan Behnke works in the lab tent set up on the Siberian tundra.*
LEFT, BOTTOM: *Blaize Denfeld (left) and her advisor, Karen Frey, filter water samples on the barge, Cherskiy, Siberia.*
OPPOSITE: *A juvenile bluethroat blends into the taiga ground cover near Cherskiy.*

permafrost thawing. The attendees were dumbstruck at the scale of the problem. Finally one of them asked, "So what's the good news?"

Holmes had no answer to that question. Luckily, Natali did: "Now you know!"

For scientists, the scientific adventure itself offers an antidote—up to a point, anyway—to the dismaying findings that research uncovers. It provides a motive to keep going, even to find delight amid adversity and foreboding. For a certain breed of scientist, as for the explorers of old, the rigors of the Arctic exert their own allure. "I went to the Arctic because I was interested in globally relevant biogeochemical processes," muses Natali. "I stayed because I loved the place. If it were a place I hated being in, I might not have. A nonscientific connection is important to have when you're working long hours in difficult conditions."

Jessica Dabrowski, a Polaris alumna and current PhD student at Massachusetts Institute of Technology and the nearby Woods Hole Oceanographic Institution, can attest to the Arctic's magic and to the impact Polaris can have in infusing budding scientists with it. "The project has changed what I'm doing with my life," she says. "When I came I knew I wanted to study environmental chemistry of some kind, and human impacts. But because of this project I've become obsessed with the Arctic—it's really beautiful and really vulnerable." And so she's applying her chemical tools to Arctic waters, studying the effects of stream and groundwater runoff from warming permafrost on the Arctic Ocean. "The program nurtured our love for place," says Dabrowski, "as well as our intellectual interest."

"Falling in love with a place that's in danger makes you fight harder for it," says Megan Behnke, a 2014 and 2015 Polaris student now working on her PhD at Florida State University. What's happening to the tundra is

OPPOSITE: *Rhys MacArthur takes a core sample of the active layer in the Y-K Delta.* **RIGHT:** *Polaris students set up a chamber to measure gases in the taiga forest, Siberia.*

"objectively terrifying, the way all climate change is objectively terrifying," she says. "It makes you feel powerless. But in research I'm also trying to deal with an interesting problem. It helps me get through the day by intellectualizing it as an issue. Curiosity makes it easier to not feel paralyzed and become a weeping heap on the floor. Getting to study the changes doesn't make them less frightening, but it helps me live with the fear."

So-called nonscientific connections like love of place, reinforce an essential scientific connection: the hunger to answer a question, to get to the heart of one piece of reality. "It's the paradoxical nature of the enterprise," says John Schade. "Scientists may lament and fear the implications of their findings, but they delight in the process of uncovering them. Unraveling complex hydrological, atmospheric, geochemical, and biochemical processes on the literal and figurative frontiers of human experience is heady work—especially for student scientists making what are for some their first forays into fieldwork, even as they face the prospect of much more dire climate impacts in their lifetimes than their elders will ever see.

JOSHUA REYES

A catastrophe in the tropics primed Joshua Reyes for the 2018 Polaris program, showing him how vulnerable natural and human systems are to climate-related disruption. After Hurricane María devastated Puerto Rico, Reyes, a native of the island and student at the University of Puerto Rico, worked with FEMA to help residents get the information and assistance they needed. The reports that trickled out from the island after the storm were bad enough, but he saw more impacts that did not make the national news: A famous beach at Manatí was swept away, together with the mangroves protecting it. A hazardous-waste storage tank burst, contaminating water supplies that the authorities insisted were safe. A week after the storm a dam broke, drowning people in its path. Rat populations exploded, contaminating food supplies and sickening those who had to eat them.

"It piqued my interest in climate change," Reyes says, "and in how one event can affect a whole area." He and the other student researchers saw how an event's effects can persist when they examined sites on the Y-K Delta scorched by fire years ago. "The interactions between burned and unburned areas were amazing," he says. He tracked the movement and eventual fate of organic carbon released from tundra soils into various types of water bodies—lakes, ponds, and fens, as well as pore and surface waters—using a device called a spectrophotometer to discern its chemical structure through the way it absorbed light at various wavelengths. "I knew there would be some distinctions in carbon structure and composition, but I found huge differences just a few feet apart." Some samples contained so much dissolved organic carbon it overloaded the device, and he had to dilute them to get a reading.

It was heady work. "I had no idea you could extrapolate so much from one wavelength!" he said as he completed his analysis. But did it make an Arctic scientist of him? "I don't necessarily want to continue in Arctic research," Reyes says. "With my background, which is tropical, I feel like I should study that environment. But this *has* defined what I want to do—to keep working in climate science."

"It piqued my interest in climate change and in how one event can affect a whole area."

"They're a little surprised at some of the basic things we don't know yet," he adds. "They seem oddly excited. They view this looming disaster as a problem but also a challenge."

The next generation's enthusiasm reenergizes veterans like Schade. Scientists are susceptible to burnout just like anyone else; the nature of their work—exacting, unpredictable, strewn with pitfalls and hurdles, crashing against walls of public indifference and incomprehension—may expose them even more to it. The students, by contrast, are, in Sue Natali's words, "not beaten down. They feel positive. For them, there are no limits. They bring that spirit to science and to changing the world and making things better. I look at them and think, 'You're convinced it can happen? I'm convinced it can happen!'"

Natali and Holmes have drawn special inspiration from Native American and Alaskan students in the Polaris Project. "They more than any others say explicitly, 'I'm doing this because I want to share this, to bring information to my community and make it better,'" notes Natali.

"My mother's people were from here, my father's people were from here," Holmes has heard several of them say. "It boggles my mind to think about people who've lived in the same place for so long, hoping future generations will live there too. I'm envious of people who have that strong connection to a place on the earth and also to their ancestors who come from the same place on the earth." This connection even informs their knowledge of the changing climate. "They can talk to their parents and grandparents, who talk to *their* parents and grandparents. It's not just about reading a paper for them. In their own family history they can see and hear about the changes."

Jordan Jimmie, a member of the 2017 and 2018 Polaris teams, pays the connection forward, drawing from his Diné (Navajo) culture a principle, called "seven generations," that has deeply impressed Holmes and Natali: Don't just think ahead to how your actions or inaction today will affect your children and grandchildren. Consider the effects of today's decisions on seven generations to come.

Back from the Brink

Unlike Arctic researchers and climate scientists, most citizens can't transmute their anxiety and alarm into the curiosity that drives research. The prospect of upheaval in the basic systems that make our planet verdant, fruitful, extravagantly rich in life and variety, and, very simply, *habitable* is daunting at the least. Scientists warn not just of the dire effects to come if we don't wean ourselves off fossil fuels but of the effects of greenhouse gases already pumped into the atmosphere, there to reside for hundreds, in some cases thousands, of years; as the oceanographer Burke Hales puts it, "We've mailed ourselves a package and now we can't call it back." Some politicians speak piously of the need to address climate change but nudge it down the list of policy priorities, below the hot-button social issues that impassion various segments of their constituencies, while others stonewall any talk of climate or brazenly deny any human impact on it.

Meanwhile, gas is cheap and life goes on. Global warming? Everyone got excited about that 20 years ago, so we must have fixed it. Or we don't need to fix it. Or we can't fix it, so why worry about it? Human nature is wired to deal with immediate, urgent threats, not long-term problems requiring comprehensive, cooperative action and short-term sacrifice for long-term gain.

Faced with these prospects, it's easy to take refuge in despair, to think, If the glaciers are melting and the coral reefs are dying, I just want to see them before they go. But as the journalist and indefatigable optimist Norman Cousins liked to say, "Nobody knows enough

OPPOSITE: The Amazon rainforest: lungs of the planet

to be pessimistic." Cynicism, like so many other drugs, provides temporary escape but eventually turns toxic and crippling.

Human nature is more adaptable than the doomsayers and denialists suppose; that's the secret of *Homo sapiens*' runaway success, which has brought us to our current climate plight. Change is really slow, until it's fast; societies as well as natural systems reach tipping points. They can change direction and shed attitudes and policies that seemed implacably embedded in culture in a historical blink of an eye. Consider how quickly the isolationist United States and its similarly unprepared allies mobilized to defeat the Axis powers after getting caught flatfooted in 1939–1941. Or how racial and gender equality came to be legally enshrined and widely (if imperfectly and inconsistently) accepted in a matter of decades after centuries of official discrimination and worse. The pendulum swung for marriage equality in just a few short years.

Mass mobilization and systemic change ultimately depend on concerted (i.e., political) action. But they well up first from the bottom, from millions of personal conversions, of examples set and persuasions achieved, ultimately coalescing in popular demand that politicians cannot ignore. The critical federal civil rights legislation of 1964–1965 did not come from on high; it grew out of decades of advocacy and agitation, from lonely personal protests to mass marches and boycotts. As President Franklin D. Roosevelt is famously (though perhaps apocryphally) said to have told one of those early advocates, "Okay, you've convinced me. Now go out there and bring pressure on me."

Making carbon-sensible choices in our daily lives is the first step in stepping up. Beyond the intrinsic value of the action is the little nudge this adds toward moving society as a whole. Stepping up puts a face on change, proving that it is possible and that the world as we know it won't come to an end if we embrace it. Reducing one individual's greenhouse impact makes an infinitesimal difference on the global scale. A million or billion reductions start to add up. The example they set and the market changes they drive may matter even more.

At the political level, those who understand climate change as an existential challenge need to act, and especially to vote, on that understanding. Gun rights and antiabortion advocates have something to teach on this score. Their political influence derives not from their numbers but from their fervor and focus; they demand *action*. Candidates know they will lose votes if they cross them. By contrast, concern for the climate and support for environmental protection have long been like delta floodwaters, broad but shallow. Poll respondents have routinely voiced support for the environment but placed it far below such priorities as the economy, health care, education, crime, civil rights, and immigration. Advocates must show the great mass of lip-service supporters that a rapidly warming climate will destabilize everything else—that, as the great civil rights leader James Farmer said, "if we do not save the environment, then whatever we do in civil rights will be of no meaning, because then we will have the equality of extinction."

In early 2019, as this book was nearing publication, campaigns such as Extinction Rebellion and the School Strike for the Climate and the debate over the proposed Green New Deal following a rash of severe storms, floods, droughts, and wildfires, had reinvigorated public awareness of climate change. The semiannual *Climate Change in the American Mind* survey found that most voters recognize its seriousness and nearly 40 percent say they'll judge candidates according to their commitment. But because climate is everyone's concern, it can too easily become no one's mission. Anyone can assume

that somebody else will take care of it. Most of those most immediately affected—tiny island nations facing inundation, farmers in sinking and salinifying flood-plains, hurricane-battered Caribbean islanders, Arctic peoples who see the ground collapsing beneath them—are poor and marginalized. The wealthy and powerful who profit most from the exploitation of fossil fuels are insulated from its effects.

These moneyed interests have successfully promoted the view that even if weaning ourselves off fossil fuels might be desirable in itself, it would be much too expensive, especially for those of modest means. This argument is losing resonance, however, as the costs of wind- and solar-generated electricity come down, approaching and even bettering those of dirty coal power. Economics are already driving a big shift from coal to natural gas, which produces only about half as much carbon dioxide as coal when burned but releases damaging methane during extraction and transportation.

When the unaccounted costs called "externalities" are added to the equation, the economic arguments for renewable energy technologies become compelling. And these technologies, together with mass transit and other consumption-side strategies to reduce fossil-fuel use, yield many other benefits as well.

Climate impacts aside, extracting, transporting, and burning fossil fuels can be horribly polluting. Coal mining levels mountaintops, leaches toxins, and fills streams with mine waste. Oil tankers ground and spill. Pipelines burst and leak. Fracking for gas and petroleum contaminates aquifers and even triggers earthquakes. Burning coal releases a massive load of pollutants, not just particulates, a.k.a. soot, but sulfur dioxide, mercury, cadmium, lead, thorium, and other heavy metals and nucleotides. These are estimated to cause 52,000 premature deaths a year from cancer, lung and heart and

LEFT, TOP: *Protected Brazil nut trees stand alone on soy farms in former Amazonian rainforest.*
LEFT, BOTTOM: *Rainforest yields to a soy farm in Mato Grosso, Brazil.*
OPPOSITE: *By law, a thin buffer of trees is preserved around a wetland in Mato Grosso, Brazil.*

OPPOSITE: *Smoke belches from a coal-burning power plant in Russia.*

kidney disease, and other ailments in the United States and more than half a million in China, and to sicken millions more. As Aiyu Zheng, a 2018 Polaris participant from the Chinese province of Sichuan, points out, this is a main reason her country's leaders are pushing so hard to reduce its carbon emissions, at the prodding of a citizenry fed up with pollution.

Clean energy is not without environmental concerns. Hydropower reservoirs can emit large quantities of methane as anaerobic bacteria digest inundated vegetation; designing and placing dams to minimize reservoir surface area relative to volume and to draw water from the top rather than methane-rich lower depths can reduce these emissions. Polluting by-products from solar-panel manufacture must be captured, and wind towers must be sited to minimize bird kills. These are small challenges compared to the damage fossil-fuel use causes.

Burning these fuels is the leading but not the only way humans pump carbon dioxide into the atmosphere. Fossil fuels contribute about 10 billion tonnes of carbon each year. Deforestation contributes another 1.5 billion tonnes—less than it did a decade or two ago but still a massive quantity. Deforestation doesn't just cut away pieces of the planet's fabled "lungs," its capacity to absorb and sequester carbon; it also slashes biodiversity, degrades air and downstream water quality, and causes wasteful erosion and destructive floods.

This forest loss is intimately connected with agriculture, which drives about three-quarters of tropical deforestation. Unlike forests that are logged and then replanted or merely left alone, those that get burned or razed to clear land for crops and livestock do not regenerate. And those uses continue to compound the climate impact.

Whether forest land is converted to pasture or cropland, it ultimately goes to raise livestock: new supplies of feed crops, primarily soy and corn, are continually needed to feed the growing demand for meat and dairy products as societies become more affluent. According to the UN Food and Agriculture Organization (FAO), 33 percent of the world's cropland is used to grow feed, not food. We could feed the planet's current human inhabitants and the two billion more it's expected to host by 2050 and restore millions of acres of precious forest—if everyone miraculously adopted a plant-based diet.

That's just one of several big ways that eating less meat helps the climate. As discussed earlier, methane has a greenhouse impact many times that of carbon dioxide, lending urgency to the Polaris Project's efforts to determine how much of the carbon trapped in Arctic permafrost will escape as CO_2 and how much as CH_4. And a surprising share of the methane in the atmosphere comes from the guts of ruminants—in particular, cattle and sheep. These animals host bacteria that break down the plant cellulose that mammals are otherwise unable to digest—and, in the way of anaerobic bacteria, release methane.

NASA has calculated that livestock are the largest source of human-caused methane emissions worldwide, nearly matching the emissions from wetlands, the largest natural source. (Fossil fuels are next largest, followed by rice farming, mainly in Asia; flooded rice paddies act as methanogenic wetlands. In the United States, livestock waste, especially from pigs, is another leading source, as are landfills, natural gas leaks, and coal mining.) The US Environmental Protection Agency estimates that since 1750, anthropogenic methane has had three-fifths as much greenhouse impact as human-produced carbon dioxide, which has gotten much more public and policy attention.

Raising animals for slaughter, milk, and eggs generates greenhouse gases in other ways as well, from

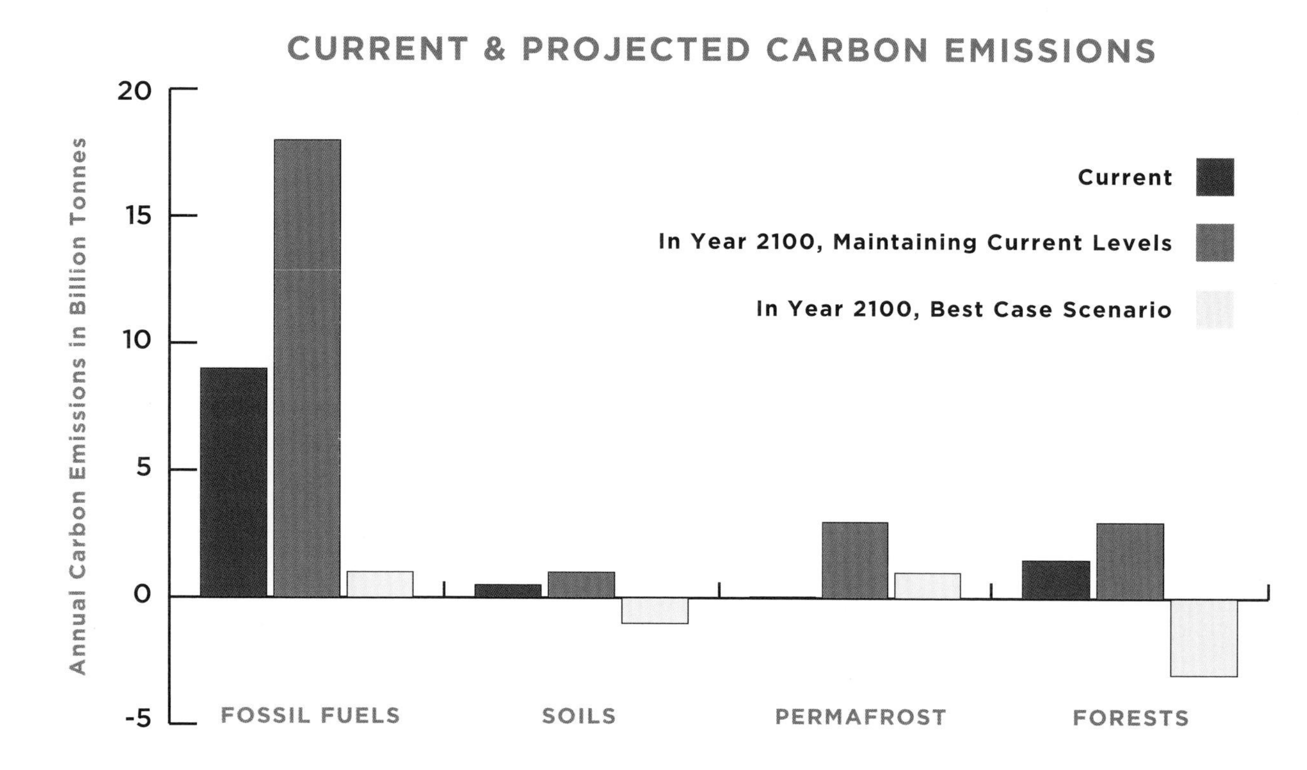

Careful management can shift soils and forests from being sources of carbon to the atmosphere to being carbon sinks, thus helping to mitigate climate change. Conversely, permafrost is likely to lose carbon to the atmosphere over the coming decades under any foreseeable scenario, but the magnitude of the carbon loss depends on how aggressively we control greenhouse gas emissions from other sources.

growing their feed to transporting and processing the final product. Both their excrement and the fertilizer used to grow their feed emit nitrous oxide, a greenhouse gas with, molecule for molecule, nearly 300 times carbon dioxide's global warming potential (GWP, a standard measure of climate impact).

As a result, the FAO attributes 14 percent of greenhouse emissions to livestock production. This impact is not equally distributed: according to a comprehensive 2018 report in *Science*, beef has about one-and-a-half to four times the average greenhouse impact per gram of protein of cheese, depending on whether it comes from cattle raised for dairy or for meat. It has two to six times the impact of pork, three to nine times that of farmed fish and chicken, four to twelve times that of eggs—and up to 150 times as much impact as beans, peas, nuts, and other plant protein sources.

For individuals seeking to reduce their climate impacts, those ratios may offer low-hanging fruit, to mix gastronomic metaphors. Eating down the meat chain (and perhaps switching from rice to any number of grains that aren't grown in methane-making paddies) may be easier for many consumers than making big changes in housing and transportation, such as forgoing driving, air conditioning, and far-off vacations.

At home and on the roads as well, there are many ways to reduce our greenhouse inputs without major public infrastructure investments: insulating and weatherizing our houses; installing solar panels, heat pumps, high-efficiency furnaces, and on-demand water heaters; switching out incandescent light bulbs; disconnecting "vampire" appliances—all these reduce home energy use. So do neglected old-time practices such as hanging wash on the line and installing fans and strategic vents instead of air conditioning—at zero or minimal expense.

Taking transit, biking, walking, and carpooling, consolidating errands and reducing trips, telecommuting and teleconferencing, taking the train instead of a plane, driving hybrids, gas misers, or electric vehicles—all these reduce transportation impacts. All are familiar to consumers today, as are the standard waste-reduction strategies: reduce, reuse, recycle, and compost, avoid excess packaging, use cloth shopping bags, don't buy what you won't eat, tell the bartender to hold the straw. All these measures are valuable to varying degrees, not just for their immediate effects but for the examples they set and the market incentives they help create. Taking transit demonstrates demand for transit, which encourages the development of more and better transit, which induces more people to take transit—a *positive* positive feedback loop.

All Together Now

But it will take collective as well as voluntary individual action to slow climate change to a pace human and natural systems can adapt to. Technology in itself will not; we cannot wait passively for human ingenuity and future inventions to save us, as some in the complacent defer-and-delay school of climate response advocate. As Nathaniel Rich writes in his 2018 investigation "Losing the Earth: The Decade We Almost Stopped Climate Change" in the *New York Times Magazine*, "New technologies had not solved the clean-air and clean-water crises of the 1970s. Activism and organization, leading to robust government regulation, had."

Regulation need not be solely coercive, however; a combination of rules and incentives, reinforced by popular sentiment, may work where neither would alone. In past decades, some people voluntarily sorted out their cans, jars, and newspapers and hauled them to recycling stations, even though they gained only the satisfaction of doing so. Most did not. Putting a price on garbage added a financial incentive to recycle instead of dump. Education and peer-group pressure did the rest; recycling is now habit in most of the United States.

Likewise, putting a price on carbon emissions through a tax or cap-and-trade plan would give everyone along the chain, from energy companies to manufacturers and consumers, incentives to reduce fossil-fuel use and shift the cost burden from the general public to actual users. A carbon tax or fee offers advantages of simplicity, clarity, and low expense to implement and administer. By rewarding rather than dictating climate-protecting measures by the private sector, it can encourage innovation and make it possible to lighten regulatory burdens. As former Federal Reserve chair Janet Yellen has said, this strategy is "absolutely standard textbook economics." It's drawn strong support from many conservative political figures, including James A. Baker III, George Schultz, and Trent Lott, who no longer have to worry about political outcomes.

Cap-and-trade, which allocates emission allowances that can then be traded on the open market, likewise incentivizes rather than dictates carbon-thrifty measures. It makes it possible to set specific emission-reduction targets rather than estimating the impact of tax incentives, and it tends to provoke less opposition than an outright tax. Cap-and-trade, which proved effective in the successful response to acid rain in the 1980s, works best with large industrial emitters such as power and cement plants but would be too costly and

OPPOSITE: *Towers of carbon-rich soil on the banks of the Kolyma River remain after the ice between them has melted.*
RIGHT: *A stream trickles through thawing permafrost at Duvannyi Yar, Siberia.*

unwieldy to apply to transportation or home heating. Each approach has its advocates. Either is much better than nothing.

But both approaches require political will and a consensus that has been sorely lacking. With basic respect for science—to say nothing of political will—in short supply at the highest levels of power of late, change must come from the bottom. What can we citizens do to overcome inertia and outright resistance to climate action? Educate ourselves, for starters, and then translate what we learn into action. We need to talk to people outside our comfort and belief zones; rather than dismissing climate skeptics and denialists as hopelessly benighted, we need to understand why they see the world as they do and find the points of agreement on which persuasion can build.

And as every parent eventually learns, setting an example can do more than lectures, pleas, or threats to encourage good behavior. On the road to Woods Hole, Massachusetts, a single wind turbine stands proudly above the trees. It is the first sign that visitors are approaching the Woods Hole Research Center. Beneath the turbine tower, beside the center's driveway, stands a large array of photovoltaic panels, recently installed.

"I've been obsessed with permafrost forever," says Ann McElvein, a Kentucky native studying at University of California Berkeley and a member of the 2018 Polaris expedition. "Maybe it was something I read, maybe in *National Geographic*." Psyched though she was for Arctic science, however, she arrived as a neophyte at the sort of hands-on, boots-in-the-muck research that Polaris entails; her studies emphasized more high-level, wide-view approaches such as GIS (satellite-based geospatial information systems), which she minored in. "I really, really like GIS. This was my first time doing fieldwork, so I learned the difference between big data and real research."

McElvein combined the macro and micro views by using a classification algorithm in the Google Earth Engine to sort high-resolution imagery of the Polaris study area into six landscape categories—burned-over and unburned dry uplands, wet lowlands, and lakes—and taking three to six sediment core samples from each. She then used a technology called infrared spectroscopy to analyze these samples for carbon content and the degree to which organic materials had decomposed (the precondition for carbon dioxide and methane releases). The findings for each landscape class may then be used to assess the vulnerability of carbon stores across the Y-K Delta.

Getting even a score or so of representative core samples proved a struggle. "When you go into the field you are sweating and crying for every data point. I had to wade across lakes and get eaten alive by mosquitoes to get one soil core and one tiny bit of information. I'm so amazed to read papers that have years and years of data. So many lives go into one paper."

"I really, really like GIS. This was my first time doing fieldwork, so I learned the difference between big data and real research."

But the experience was a revelation. "You can read a book about permafrost," says Paul Mann, "but until you push a probe in and it only goes this far and hits something like rock, or put your hand down in what looks like normal soil and hit solid ice, you'll never appreciate it. It's thrilling."

When the Polaris team finally arrived at the field camp, McElvein and another student were the first to drill a core sample from the permafrost. To anyone not aware of that sample's import and promise, it would look like just a dull brown cylinder of hardened dirt. When she saw it, she couldn't help crying. Even Max Holmes, who's seen many, many core samples taken, was moved: "I almost started to cry too."

"When you go into the field you are sweating and crying for every data point. I had to wade across lakes and get eaten alive by mosquitoes to get one soil core and one tiny bit of information."

Two working arrays sit on the center's roof. For years, the roof arrays and wind turbine supplied about half WHRC's power. The new solar capacity, together with energy-saving improvements, have now made it carbon-neutral.

Installations like these help create a climate of possibilities in another sense as well. When the wind tower went up, some neighbors complained. WHRC's engineers adjusted it for wind speed and direction, reducing the noise it produced. Complaints stopped, and the tower is now a local landmark, visible testimony to an alternate energy future. Fifteen years ago, when WHRC went solar, photovoltaic panels were nearly unknown around Woods Hole and nearby Falmouth. Now they are familiar rooftop fixtures on Cape Cod, like seagull weather vanes.

You Get What You Need

It's harder to spot such hopeful signs on the tundra. Polaris veterans like Paul Mann returned from their 2018 expedition to the Y-K Delta, amid a season of record Arctic warming, feeling sobered. "I saw a lot more dried lakes this year," Mann muses. "Places I'd visited before had changed dramatically. There's a growing sense of how *precarious* this region is." Sue Natali noticed many more cracks in the ground, formed as frozen soil thaws and dries, than in previous years. In the area around the Polaris camp, the average soil temperature one meter down increased by about 0.2° Celsius (0.36° Fahrenheit) from summer 2017 to summer 2018. At just a tenth of a degree below the freezing point, the ground is now precariously close to thawing.

For the Native residents of the tundra, in Alaska as in Siberia, these changes are matters of daily struggle, not seasonal investigation. Sitting around a table in WHRC's canteen, several of the 2018 student researchers recalled what they'd learned from the people of the

OPPOSITE: *Common cottongrass* (Eriophorum angustifolium) *grows along the marshy banks of the Panteleikha River, Siberia.*
RIGHT: *Polaris faculty member Karen Frey takes a water sample from a clear mountain stream near the Kolyma River Delta.*

Yukon-Kuskokwim Delta about life in the thaw zone. "It was startling, but it was also refreshing," mused Natalie Baillargeon, "because here [in the Lower 48] we're so used to climate change being a controversial issue when it shouldn't be, but there it's a fact of life. Our work here is so important because if we're able to spread the message that this is happening and this is why we're doing our research, then that startled feeling that I felt can be shared among the general public."

"I tend to think of a song by Lauryn Hill," Jordan Jimmie interjected. "Before one of her unplugged concerts she said, 'Fantasy is what people want. Reality is what people need.' Fantasy is the privileged world we live in. The reality is, outside our borders, people are being displaced because of extreme weather events. One day we're going to have that one catastrophic event, and we're going to be the refugees."

Can nothing short of disaster shake us out of our collective complacency? Knowledge alone does not solve a problem, but it is the indispensable base on which solutions are built. As Sue Natali says, "Now you know." Better to get to work now than to wish we had later.

Nikita Zimov drives the "water roads" (the only available route through this roadless wilderness) to Pleistocene Park.

Larch trees lean over a thermokarst lake formed by thawing permafrost near Cherskiy, Siberia.

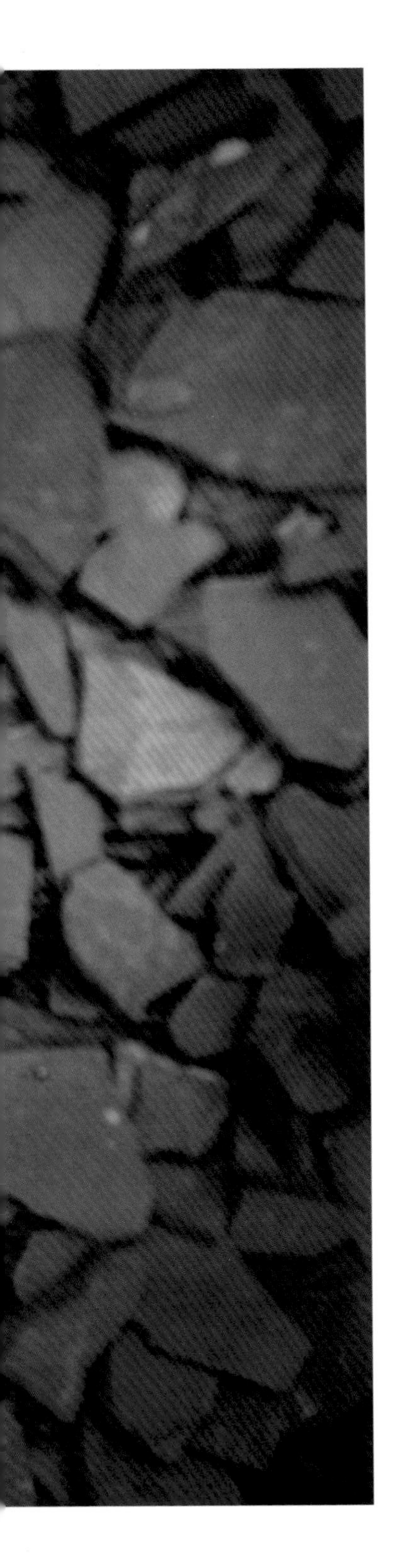

EPILOGUE

By Theodore Roosevelt IV

***THE BIG THAW* MAKES IT** clear that the planet has reached a crossroads with respect to climate change. The road we take will make all the difference. The average global temperature has increased by almost 1° Celsius since the beginning of the Industrial Revolution. Based on what they know about the amount of greenhouse gases (GHG) that have accumulated in the atmosphere,

Arctic sandwort (Minuartia arctica) *grows on a windy, shale-covered slope overlooking the Kolyma River in Siberia.*

most scientists think it will be virtually impossible to limit global warming to an increase of 1.5° Celsius—the objective of the Paris Climate Agreement. Indeed, many experts believe that without dramatic decreases in our use of fossil fuels, the increase will exceed 2° Celsius.

Seemingly small changes in the average global temperature have a major impact on planet Earth. For example, a temperature change of only 5° Celsius separates the planet we know today from the last ice age twenty thousand years ago when most of the northern hemisphere was covered in thick sheets of ice. The changes we are already seeing, which entail material risk for the planet and all the species that live on it—including *Homo sapiens*, have been accurately predicted by climate scientists for almost half a century: rising seas, longer and more frequent droughts, increased ocean acidification, more climate disruptions, wildfires, and transnational migrations that add to geopolitical instability.

As a result of work by the Woods Hole Research Center (WHRC) and the broader scientific community, we now know that the Arctic permafrost contains more carbon than all of Earth's forests. In fact, it is estimated that the entire permafrost region contains between 1,200 to 1,850 billion tonnes of carbon. The midrange of 1,500 billion tonnes is larger than the 1,200 billion tonnes of all the remaining accessible fossil fuels and almost twice the 850 billion tonnes of carbon in the atmosphere. What we don't know is how much carbon is currently being released from the permafrost. As the magnitude of the emissions from the permafrost region becomes known, a corresponding decrease in GHG emissions from all other sources will be required to ensure the GHG "budget" established by the Paris Climate Agreement is not exceeded. This means that the time available for meeting the climate challenge is shorter than current estimates, which do not yet take into account the release of carbon from the permafrost. Expressed another way, we are closer than we think to the tipping point at which it will be too late to successfully address this existential challenge.

There is good news: the cost of renewable energy has fallen dramatically. As Bloomberg New Energy Finance (BNEF) reports, the cost of renewable energy from onshore wind, offshore wind, and solar photovoltaic has fallen by 49 percent, 56 percent, and 84 percent respectively since 2010 as of this writing. The cost of lithium-ion battery storage has dropped 76 percent since 2012. The BNEF report continues, "batteries co-located with wind or solar projects are starting to compete in many markets with coal and gas fired generation for dispatchable power. It is no longer true that wind and solar are more expensive than coal or natural gas and in many markets wind and solar even with storage costs included compete with coal and natural gas."

We owe the scientists from WHRC and those with whom they collaborate an immense debt of gratitude. Their work provides us with essential information we need about the greenhouse gases being emitted from the permafrost belt in order to create successful policies to meet the climate change challenge. Prudent risk management dictates that, with unflinching determination, each of us must do everything possible to ensure climate policies are aggressively pursued to avoid catastrophe. Acting as responsible stewards for future generations is simply a moral responsibility that we must meet.

OPPOSITE: *Common mare's-tail* (Hippuris vulgaris), *looking like a miniature coniferous forest, grows in Siberian bogs.*

ACKNOWLEDGMENTS

When Chris Linder and I first pondered a book project several years ago, we knew we had an important story to tell and powerful photos to help tell it. But the pathway from idea to reality was unclear, so perhaps it should not be surprising that we had some false starts along the way. But then we connected with Helen Cherullo, Janet Kimball, and Braided River, and through them we partnered with the journalist Eric Scigliano. The path has still been challenging, but we're thrilled to have reached the end and to be able to get the book into the hands of our intended audience—you!

The work that is the focus of this book could not have happened without a tremendous number of people: scientists in the US and Russia, students from both of those countries and more, along with the funders who made the research—and this book—possible. On behalf of the entire Polaris team, particularly Dr. Susan Natali and Dr. John Schade who played major roles in making this book a reality, my heartfelt thanks goes to all of them.

Special thanks also to my wife, Gabby, and my kids, Nate and Sophie, who didn't directly experience the adventure and excitement of the Arctic expeditions but who did have to carry my load when I was away from home for weeks at a time. Thanks to all of them and to you for holding this book in your hands and learning more about this remarkable planet we call home.

—Dr. Robert Max Holmes

I am immensely grateful to the many people who helped make this book possible. The roots of this project date back to 2008, when Max Holmes invited me to join him on a winter expedition to Zhigansk, Siberia. That life-changing experience led to seven more expeditions with the Polaris Project. Thank you, Max, and the Woods Hole Research Center, for believing in the power of photography to communicate science stories.

Thanks to all of the scientists, postdocs, and students who have so generously taken time to explain their work and tolerated having a camera stuck in their faces . . . but even more than that, thanks for the camaraderie and laughs. You remind me that the most important thing about science is the scientists themselves—curious people in search of truth.

Thank you to the Zimov and Davydov families, who kept me fed, safe, and entertained in the Siberian Arctic.

Thank you to Helen Cherullo at Braided River for your support and thoughtful advice over the years. Thanks also to Janet Kimball, Jen Grable, and the entire staff at Mountaineers Books and Braided River for turning ten years of photographs and stories into these engaging, beautiful pages.

Thank you to my wife, Meghan, and my kids, Eva and Ian, who have endured the hardships of separation while I was in the field. The secret notes you put in my luggage kept me smiling, even when I was covered with mosquitoes or up to my knees in permafrost quicksand.

—Chris Linder

WITH APPRECIATION

WOODS HOLE RESEARCH CENTER WOULD LIKE TO THANK THE FOLLOWING SPONSORS FOR THEIR GENEROUS SUPPORT:

Benefactor

Ruth McCormick Tankersley Charitable Trust

Patron

Brabson Library and Educational Foundation
Furthermore: a program of the J. M. Kaplan Fund

Subscriber

Molly N. Cornell
The Dorothea Royer Endicott Foundation
Tom, Ayla, Sunny and Maia Fudala
Nora Greenglass
Mary and Harry Hintlian
Robert and Pamela Pelletreau
Amy and Jay Regan
Scott Wayne

In memory of

Craig B. Speiser, grandson of Lawrence and Lois Woodwell

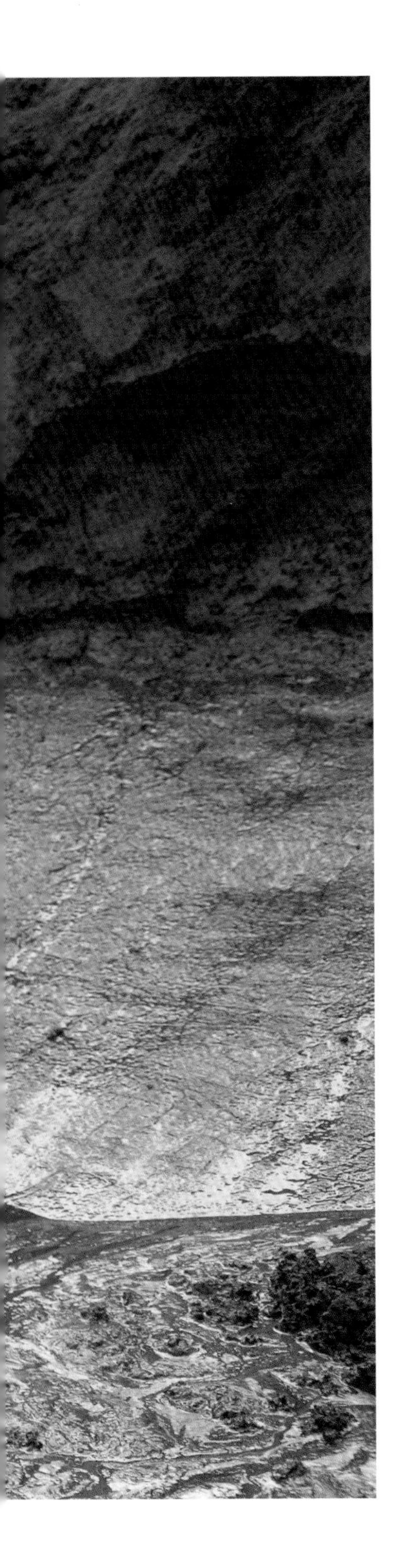

NOTES FROM THE FIELD

By Photographer Chris Linder

MUD AND INVISIBLE GAS. I'M not sure there are two less photogenic (or difficult!) things to photograph, but they are central characters in this story. Permafrost is best appreciated through senses other than sight; it feels like frozen cement in your hand when you chisel it out of the ground, and when it thaws, it oozes like chocolate pudding. It has a rich, earthy smell, like freshly turned

Making an engaging photograph of frozen mud isn't easy. Chris Linder sets up his tripod at the Duvannyi Yar riverbank exposure in Siberia. (Photo by Max Wilbert)

soil in a farmer's field. But to the eye, it doesn't look a whole lot different from the mud pies my kids made as toddlers. Methane and carbon dioxide gases are the by-products of microbes that eat the carbon contained in permafrost, but they are completely invisible . . . unless you watch them bubble up from the bottom of an Arctic lake.

After my first frustrating summer photographing this project, I realized I needed to start thinking outside the box. I invested in two new pieces of equipment: an underwater housing for my camera and a wetsuit—because Arctic lakes and rivers, even in summer, are *cold*. This opened up a whole new perspective for me. Now I could be *in the river* as researchers leaned over the sides of the aluminum speedboats and collected water samples. I could capture those huge clouds of methane I saw escaping from the bottom of the lakes. As an added bonus, I found the wetsuit's neoprene was thick enough to repel even the most persistent mosquito.

The next toughest part of my job was just being in the right place at the right time. Summer in the Arctic means twenty-four hours of daylight, with the best light happening when most sane people are sleeping. Capturing both the science team in action and the peak landscape light meant eighteen-hour workdays for me. My daily routine went something like this: follow researchers from 8:00 a.m. to 8:00 p.m., sleep for a few hours, shoot landscapes from 11:00 p.m. to 3:00 a.m., sleep for a few more hours, and crawl back out of bed at 7:00 a.m.

There's something nourishing about Arctic light, though. It washes away weariness nearly as well as a good night's sleep. Sitting on a bluff over the Kolyma River watching the sun scoot sideways between distant peaks, I couldn't imagine missing one second of every three-hour sunset or sunrise.

From the Midwest to Siberia

When I was a kid growing up in Virginia and Wisconsin, I spent most of my time covered in mud in pursuit of snakes, turtles, and newts. I'll never forget the look on my mom's face when I proudly showed her a three-foot-long garter snake or my own horror at the number of leeches I had to remove after a swim in the local river. My hero was Jacques Cousteau, and I was enthralled by his television program, *The Undersea World of Jacques Cousteau*. When my dad and I paddled our Grumman canoe down the Bark River, I imagined I was exploring hidden worlds and discovering new and exciting species. Though I had only seen the ocean a handful of times by the time I graduated from high school, I was convinced my future was tied to the sea.

I attended the US Naval Academy where I studied oceanography and worked on my own research project using sonar to image oyster beds in Chesapeake Bay. I continued my studies at the Massachusetts Institute of Technology and the Woods Hole Oceanographic Institution (WHOI) where I earned my master's degree and published my research on the New England shelf-break front, a water mass boundary separating cold, fresh coastal waters from the warm, salty Gulf Stream. After finishing my academic training, I returned to the Navy, where I served as a junior officer in their Meteorology and Oceanography program. I spent the majority of my service in Rota, Spain, forecasting the weather for the Mediterranean fleet.

Just before moving to Spain, I got my first Nikon camera. Every weekend that I was off duty, I took pictures. I documented colorful spring fairs, tiny mountain villages, and Andalucía's extensive network of parks. What started as a hobby quickly became an obsession, and my cabinets began to fill with pages of Fuji Velvia color slides.

Finding a better angle (or just trying to escape the mosquitoes) in a larch tree (Photo by Brian Kantor)

When my service commitment ended, I returned to WHOI as a researcher, working with my former thesis advisor on projects from Cape Hatteras to Belize to Taiwan. When I went to sea on research vessels, I brought my camera along. Between watch shifts, I photographed, not just the beautiful sunsets, but the people doing the work—scientists collecting data, the ship's crew driving the ship, even the cook preparing the meals. With my photographs, I could explain cutting-edge science in a visual way that people of any age could appreciate. As my photographic skills grew, scientists at WHOI began to hire me to document their projects. In 2002 I created my first expedition blog. In addition to the photography, I wrote dispatches, answered questions from schoolkids, and even coded the HTML for a project studying the Chukchi and Beaufort Seas off the coast of Alaska. After that experience, I was hooked. The prospect of combining my passions for oceanography and photography was exhilarating, and within a few years I had transitioned to full-time freelance scientific expedition photography.

After completing a number of expeditions to both the Arctic and Antarctica, a program manager at the National Science Foundation encouraged me to talk with Dr. Max Holmes from the Woods Hole Research Center (WHRC). Max, a river chemist specializing in the Arctic, had just been funded to create a new field program for undergraduates in Siberia. Max is the kind of person who fills a room—his enthusiasm and complete devotion to his work are infectious. By the time he finished

OPPOSITE: *The sun sets behind the banks of the Kolyma River, Siberia.*
RIGHT, TOP: *Delicate flowers like this boreal Indian paintbrush* (Castilleja pallida) *bloom in the summer on the Arctic tundra.*
RIGHT, BOTTOM: *A mosquito rests on a bloom of smart weed* (Bistorta elliptica).

explaining the Polaris Project to me in his office, I had agreed to join the team as the expedition photographer. Little did I know that over the next ten years, I would be making seven trips to the Siberian and Alaskan Arctic to photograph the Polaris Project.

On Taiga and Tundra

It's midnight, and the sun has just started to tiptoe across the tops of the larch trees on the western bank of the Kolyma River. The horizon glows orange, fading to a deep purple in the clear sky above. The river is wide here, and the orderly line of treetops on the far bank, tiny in the distance, looks like the teeth of a long zipper. On the deck of our floating dormitory/laboratory (a repurposed Soviet-era barge), undergraduate students, scientists, and our Russian hosts chat excitedly and snap photos in the warm sunset light. It's the start of a Polaris research experience: we are headed north, toward the Arctic Ocean, for an expedition to study the Arctic tundra.

As we motor down the mighty Kolyma, the taiga forest rolls endlessly by. The sun never quite makes it behind the trees. After dipping teasingly low, it jumps right back up into the sky, filling the barge with an orange glow. By the time breakfast—a hot, salty, buttery gruel and homemade bread made by our Russian cook—is served eight hours later, the landscape finally shows signs of change.

The thick stands of larch trees thin and shrink. Soon, only a few twisted and ancient trees barely taller than a toddler can be seen among the rolling green expanse of the Arctic tundra. The *tundra* . . . I had expected a bleak, featureless environment sculpted by the cold and wind and harboring only a handful of plant and animal species.

But when we finally anchor the barge on a sandbar and bound up the nearby bluff, a new panorama in

miniature is revealed at our feet. On the ridges of the gently undulating landscape, dwarf willow and birch shrubs only a few inches high form a spongy green mat. Tiny blossoms of white and fragrant Labrador tea (the leaves, when crushed, smell like Christmas trees), blue forget-me-nots, delicate white and pink Indian paintbrush, and purple thyme dot the green like stars. Descending into the shallowly sloping valleys where water pools and trickles in tiny rills on top of the permafrost, foot-tall tussock mounds topped with the white balls of cottongrass dance in the breeze. Farther in the distance, abutting the banks of the river, thousands of shallow pools and floating vegetation form an intricate mosaic. This is a landscape defined by extremes in scale: endless grandeur and minute detail.

LEFT, TOP: *Chasing scientists across the Siberian tundra* (Photo by Max Wilbert) **LEFT, BOTTOM:** *The hardest and best photo shoot of my life—documenting Evenki herders working with their reindeer* (Photo by Max Holmes) **OPPOSITE:** *The sun shines through the cottongrass on a windy day near Cherskiy.*

Arctic Myths

When I tell people I'm headed to the Arctic for the summer, I get some predictable responses like "Oh, you're going to freeze" or "Watch out for polar bears." And I'm not surprised; the Arctic to most people conjures up images of a vast, frozen wasteland inhabited by polar bears (and many assume incorrectly, penguins). Arctic penguins, it turns out, are only one of several misconceptions that people have about this region. Here are a few more:

Myth #1: The Arctic is cold all the time.

It's true that in the winter it can get bone-chillingly cold in the Arctic. That's how permafrost is formed. However, in the summer, particularly in inland regions like central Siberia, the temperature can exceed 32° Celsius (90° Fahrenheit) if the wind blows from the south. The next day, it can snow. Near the coast, the Arctic Ocean modulates extremes in temperature—it is neither as cold nor as warm as the interior regions.

OPPOSITE: *Riding bareback and using a stick for balance, a young Evenki boy drives his reindeer across a frozen lake north of Zhigansk, Siberia.*
RIGHT: *The one that (almost) got away with my blood*

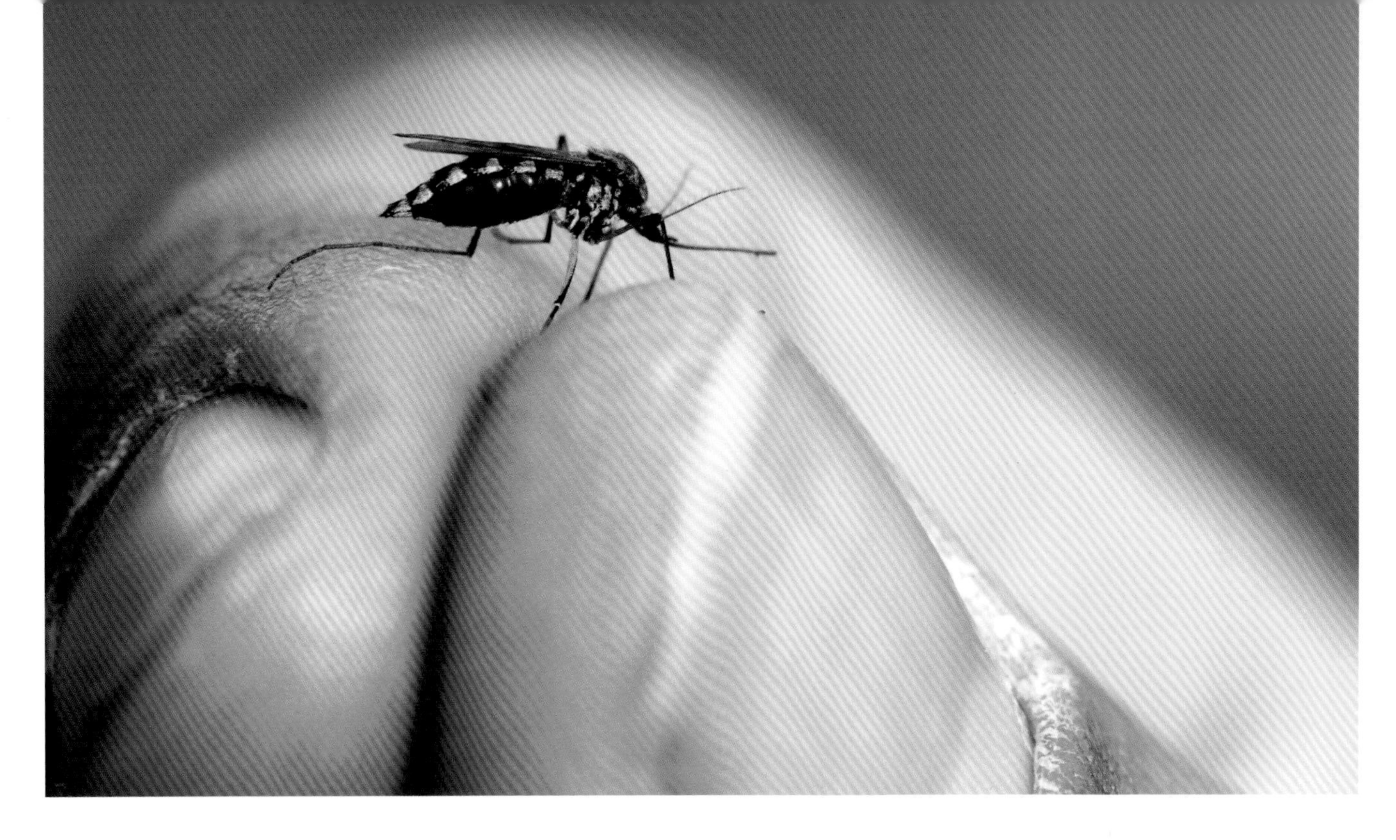

Myth #2: The ice-covered Arctic Ocean is so cold that it doesn't support marine life.
Sea ice cover in the Arctic is seasonal; in the summer, only part of the Arctic Ocean is covered by ice (and that amount is less and less every year now), while in the winter, the Arctic Ocean is completely ice covered. In the spring, when the ice retreats under the onslaught of twenty-four hours of sunlight, tiny marine plants bloom in the Arctic Ocean and feed one of the richest hot spots of marine life on the planet, including fish, seals, walrus, whales, and of course polar bears.

Myth #3: The land in the Arctic is uninhabited and devoid of life.
Four million people make their permanent homes in the Arctic (as opposed to zero in Antarctica). The taiga, or boreal, forest is the world's largest land biome. Over a quarter of the world's trees are found in this seemingly unending belt of conifers that stretches across northern Russia, Scandinavia, Canada, and Alaska. Moose, caribou, musk ox, great gray owls, and brown bears live here. In the summer, the tundra is a nursery for hundreds of species of birds that migrate north to raise their chicks. And then there are the mosquitoes . . .

Battling the Bugs

> "If you put your hand out, and you catch ten mosquitoes, *then* there are mosquitoes. Otherwise, there are *no* mosquitoes."
>
> —Sergey Zimov, director, Northeast Science Station, Cherskiy, Siberia

My first evening in Cherskiy, Siberia, almost broke me. Clad head to toe in bug-thwarting mesh, thick gloves, a Gore-Tex jacket, and knee-high boots, I confidently strode off into the boreal forest in search of the warm, delicious low-angle light that photographers live for. At 67° north latitude in July, that happens in the wee hours of the night.

As soon as I stepped out into the calm, cool evening air, the assault on my defenses began. I'll never forget the sound. Hundreds, perhaps thousands of mosquitoes were hurling themselves at my body in a desperate effort to find an exposed patch of flesh. The pitter-patter of their bodies crashing into the hood of my parka sounded like raindrops.

Confident that my defenses were holding the tidal wave of bugs at bay, I bushwhacked through the forest to what would become one of my favorite haunts: a nearby lake the locals called Schuch'ye. The mosquitoes followed, and as I moved through the forest, I stirred up more, until a veritable tornado of mosquitoes was swirling around me. Every time I paused to set up my tripod, their attacks would intensify to a fever pitch. Photobombing mosquitoes marred every shot. I struggled to evaluate my composition and precise focus while peering through the dark mesh bug net. I cursed my attackers and flailed around a bit, which of course did absolutely nothing. And on that first evening, when I came back to the station with zero usable photos, I will admit that I questioned if I could survive a month like this.

But I did. Every day got a little easier. It wasn't as much of a physical challenge as a mental one. I learned how to filter out the sound of their angry buzzing and the sight of hundreds of them covering my jacket and pants. And after a few weeks, like Sergey, I barely noticed they were there.

Hope for the Future

At the end of my first summer in Siberia, I remember having dinner in the city of Yakutsk with the science team, and one of the faculty members asked me if I was going to come back the next year. Without hesitation, I said that nothing could entice me to come back to that buggy swamp.

But over the next year, as I spent hours, days, and weeks editing interviews that I had conducted with the undergraduates and producing short multimedia videos for the Polaris website, the consequence of this project started to sink in. In just four short weeks, Polaris had clearly made a tremendous impact on the students. The process of designing their own field research and getting answers to big questions about climate change had literally changed their life trajectories. When I found out that four of the students had applied to come back in 2010 to continue their research and mentor the new students, I knew I had to come back as well and continue to document this critical juncture in their lives.

Climate change and its many global repercussions will be their generation's challenge to solve. Just being around these undergrads, watching as they come to grips with that reality, and seeing them rise to the task with enthusiasm and optimism—that's what really gives me hope. And so, with my own grim determination, I have put on that head net and battled the bugs for seven seasons to tell their story.

OPPOSITE: *Dressed for battle with the mosquito hordes* (Photo by John Schade)

Nikon
D810

ABOUT THE AUTHORS

Photo by Florence Omoro

Eric Scigliano is a journalist and author with a longtime interest in climate change, the Arctic, and the alarming intersection between the two. He has been a science writer at the University of Washington and a staff writer and editor at several newspapers and magazines, and has written for *National Geographic*, *New Scientist*, *Discover*, *Harper's*, *Technology Review*, and many other national publications. Scigliano is the author of *Michelangelo's Mountain* and *Seeing the Elephant: The Ties That Bind Elephants and Humans* (originally published as *Love, War, and Circuses*) and coauthor, with Curtis Ebbesmeyer, of *Flotsametrics and the Floating World*. His writing has been featured in four photo books including, most recently, *The Wild Edge: Freedom to Roam the Pacific Coast*, from Braided River. His work has won Livingston and American Association for the Advancement of Science awards and has been included in *Best American Science Writing*.

Dr. Robert Max Holmes is deputy director and senior scientist at the Woods Hole Research Center. He is an earth system scientist who studies rivers and their watersheds and how climate change and other disturbances are impacting the cycles of water and chemicals in the environment. Dr. Holmes is also interested in the fate of the vast quantities of ancient carbon locked in permafrost in the Arctic, which may be released as permafrost thaws, exacerbating global warming. He has ongoing projects in the Russian, Canadian, and Alaskan Arctic and in the tropics in the Amazon and the Congo. Dr. Holmes is committed to engaging students in his research projects and to communicating the results and implications of his research to the public and to policy makers. He previously served as director of the National Science Foundation's Arctic System Science Program and in 2015 was elected National Fellow of the Explorers Club.

Dr. Susan Natali is an associate scientist at the Woods Hole Research Center. She is an Arctic ecologist who studies the response of Arctic ecosystems to a changing environment and the local to global impacts of these changes. Her research examines the effects of permafrost thaw on global climate and impacts of fire and landscape characteristics on permafrost vulnerability. Dr. Natali has worked extensively in remote regions of Alaska and Siberia conducting research and, as a faculty member of the Polaris Project, training the next generation of Arctic scientists. Dr. Natali is committed to creating a more inclusive scientific community and to communicating the impacts of climate change in the Arctic to policy makers and public audiences. Dr. Natali earned her BS in biology from Villanova University and her PhD in ecology and evolution from Stony Brook University.

Dr. John Schade is an Arctic scientist and educator who, through his role as the educational coordinator for the Polaris Project, is training the next generation of Arctic scientists. Dr. Schade also serves as a distinguished visiting scientist at the Woods Hole Research Center (WHRC), studying the impacts of climate change on streams and wetlands in the Arctic. Before coming to WHRC, he spent ten years on the faculty at St. Olaf College in central Minnesota, where he began developing ideas about the education of new scientists that continue to influence his work with the students and faculty of the Polaris Project. Dr. Schade earned his BS from the University of Michigan and both his MS and PhD from Arizona State University, all in biology.

Photo by Ben Masters

Theodore Roosevelt IV is a managing director in investment banking at Barclays, based in New York. Currently, he serves as chairman of the firm's Clean Tech Initiative. He started his investment banking career at Lehman Brothers in 1972.

Mr. Roosevelt is board chair of the Center for Climate and Energy Solutions (C2ES), a trustee of the Climate Reality Project, a member of the Governing Council of the Wilderness Society, and a trustee for the American Museum of Natural History.

ABOUT THE PHOTOGRAPHER

Photo by Brian Kantor

Chris Linder is a professional science and natural history photographer. A former US Navy officer and oceanographer, Linder now focuses on communicating the stories of scientists working in extreme environments. He has documented more than fifty scientific expeditions from the Congo River to Siberia and has spent over two years of his life exploring the polar regions. Linder's images have appeared in museums, books, calendars, and magazines. A solo exhibition of his photographs, *Exploring the Arctic Seafloor*, was displayed at the Field Museum in Chicago and the Carnegie Museum of Natural History in Pittsburgh. He is the author of *Science on Ice: Four Polar Expeditions* and was the lead cinematographer for the documentary film *Antarctic Edge: 70° South*. His work has been recognized with numerous awards from international photography competitions and a prestigious National Science Foundation Antarctic Artists and Writers grant. When not on assignment, he enjoys sharing his passion for photography by teaching workshops and giving presentations to audiences of all ages. Linder is a Senior Fellow in the International League of Conservation Photographers, a Fellow National in the Explorers Club, and a member of the SeaLegacy Collective.

ABOUT WOODS HOLE RESEARCH CENTER

Established in 1985, Woods Hole Research Center (WHRC) seeks to advance scientific discovery on climate change impacts and solutions.

WHRC was founded by ecologist Dr. George Woodwell, who became alarmed in the early 1980s about the impending threat of climate change. Woodwell designed an organization that would conduct rigorous scientific research on climate change causes and impacts and also work to deliver that science to policy makers and decision makers.

Woodwell testified to a US Senate subcommittee on environmental protection in 1986 and said that "on the issue of climatic change there is surprising unanimity at the moment. If current trends continue, the increase in infrared absorptive gases in the atmosphere will lead over the next three to five decades to a warming of the earth that will average between 1.5 and 4.5° C."

He predicted that the effects of climate change would be "profound. They include effects on the distribution of arable land, the productivity of agriculture, the availability of water in lakes and streams, the productivity of grazing lands and fisheries, and the level of the sea. They also reach to effects on the distribution of the vegetation of the earth. . . . The issue seems especially clear in the higher latitudes where the temperature changes will be greatest and where the accumulations of carbon vulnerable to decay are high."

Woodwell's predictions—and those of the broader scientific community—have unfortunately become reality. WHRC scientists are measuring and studying how climate change effects are playing out—from the Arctic to the Amazon. These impacts include increased wildfires, a disrupted jet stream, increased drought, and unraveling Arctic ecosystems. WHRC's research helps decision makers understand how to prepare for impacts and also sheds light on potential irreversible changes to the climate system. Chief among these tipping points is thawing permafrost—the subject of this book.

WHRC has worked since its inception to make national and international decision makers aware of the latest science on climate change. Woodwell and his colleagues helped launch the United Nations Framework Convention on Climate Change and the UN's Reducing Emissions from Deforestation and Forest Degradation program. WHRC scientists have contributed to every assessment report of the Intergovernmental Panel on Climate Change. The center works with national governments around the world to help meet country goals for emissions reduction.

In an effort to incorporate climate science into societal decision-making, WHRC also works with nongovernmental groups, such as private industry, conservation organizations, and universities.

A thermokarst lake formed from thawing permafrost reflects cumulus clouds on a summer day in Siberia.

With increased awareness, there is still time to mitigate the worst impact of climate change. Some natural systems—such a permafrost—are emitting greenhouse gases as global temperatures increase, but forests and soils also have an enormous capacity to pull carbon dioxide out of the atmosphere. WHRC will work to help integrate climate science into the world's biggest decisions and to chart a path toward a stable climate system.

BRAIDED RIVER

BRAIDED RIVER, the conservation imprint of Mountaineers Books, combines photography and writing to bring a fresh perspective to key environmental issues. Our books reach beyond the printed page as we take these distinctive voices and vision to a wider audience through partnerships, media, exhibits, and multimedia events. Our goal is to build public support for environmental and social justice issues and inspire action. This work is made possible through the book sales and contributions made to Braided River, a 501(c)(3) nonprofit organization. Please visit BraidedRiver.org for more information on events, exhibits, speakers, and how to contribute to this work.

Braided River books may be purchased for corporate, educational, or other promotional sales. For special discounts and information, contact our sales department at 800.553.4453 or mbooks@mountaineersbooks.org.

THE MOUNTAINEERS, founded in 1906, is a nonprofit outdoor activity and conservation organization, whose mission is "to explore, study, preserve, and enjoy the natural beauty of the outdoors. . . . " Mountaineers Books supports this mission by publishing travel and natural history guides, instructional texts, and works on conservation and history.

Mountaineers Books
1001 SW Klickitat Way, Suite 201
Seattle, WA 98134
800.553.4453
www.mountaineersbooks.org

An independent nonprofit publisher since 1960

Manufactured in China on FSC®-certified paper, using soy-based ink.

For information and updates on this project, visit www.BraidedRiver.org.

First edition, 2019

Publisher: Helen Cherullo
Project Editor: Janet Kimball
Developmental Editor: Linda Gunnarson
Copyeditor: Laura Whittemore
Cover and Book Designer: Jen Grable
Cartographer: Greg Fiske, Woods Hole Research Center
Illustrator: Natalie Renier

Front cover photo: *Methane bubbles belie thawing permafrost in the Siberian Arctic;* Back cover: *Polaris students collect water samples from the clear waters of the Sukharnaya River;* Page 1: *A massive slab of thawing permafrost leans over the banks of the Kolyma River in the Siberian Arctic, dwarfing Sergey Zimov in his boat;* Page 10: *Methane bubbles up from a Siberian lake;* Pages 12–13: *Arctic cottongrass (*Eriophorum callitrix*) blooms in the Yukon-Kuskokwim Delta, Alaska;* Page 49: top, *Claire Griffin;* bottom, *Rivers and lakes in the Y-K Delta;* Page 64: top, *Travis Drake;* middle, *A researcher holds a vial of high molecular weight dissolved organic matter;* bottom, *Travis Drake and Erin Seybold in a permafrost tunnel, Yakutsk, Russia;* Page 68: top, *Mia Arvizu;* bottom, *Arctic cottongrass (*Eriophorum callitrix*), Alaska;* Page 80: top, *Kolyma River, Siberia;* bottom, *Blaize Denfeld;* Page 88: top and bottom, *Megan Behnke doing field research;* Page 112: top, *Kelly Turner;* bottom, *Scientists dig up a sample of the active layer of soil in the Y-K Delta;* Page 116: top and bottom, *Darcy Peter at work in the field;* Page 128, top, *Joshua Reyes;* bottom, *Sunset on the Kolyma River, Siberia;* Page 140: top and bottom, *Ann McElvein gathers sediment core samples;* Page 141, *A permafrost core extracted in the Y-K Delta;* Page 176, *A double rainbow arcs over a grounded barge on the Panteleikha River, Siberia.*

Library of Congress Cataloging-in-Publication Data is on file at https://lccn.loc.gov/2019015089

ISBN 978-1-68051-247-2

This book was made possible with generous support from Furthermore, a program of the J. M. Kaplan Fund.

Arctic ground squirrels are a surprising source of "rodentogenic" carbon emissions.

Mosses and plants create a quiet kaleidoscope of color around an Alaskan lake.

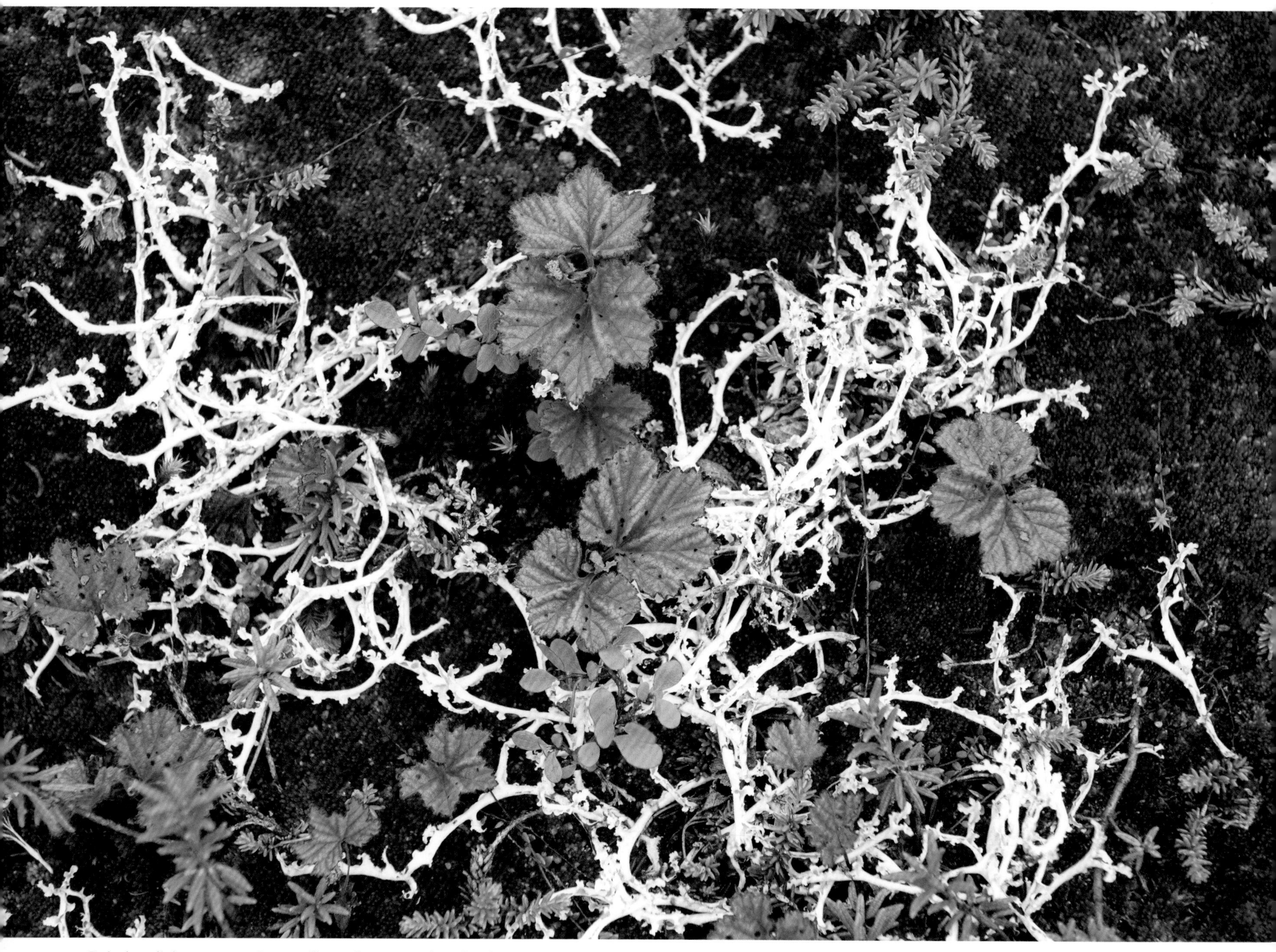

Reindeer lichen grows atop a pillow of russet-colored sphagnum moss, Alaska.

A spiderweb catches early morning dew in the taiga forest, Siberia.

A glowing green band of the aurora borealis arches over the taiga, Siberia.

Stars whirl over a now defunct satellite dish atop the Orbita lab at the Northeast Science Station, Siberia.